THE UNSOUGHT FARM

The Unsought Farm

Monica Edwards

Copyright ©Monica Edwards, 1954

First published in Great Britain 1954
by Michael Joseph Ltd

Published in Large Print 1995 by ISIS Publishing Ltd,
7 Centremead, Osney Mead, Oxford OX2 0ES,
by arrangement with Monica Edwards

All rights reserved

The moral right of the author has been asserted.

British Library Cataloguing in Publication Data
Edwards, Monica
Unsought Farm. – New ed
I. Title
942.21092

ISBN 1-85695-142-1

NORFOLK LIBRARY AND
INFORMATION SERVICE

SUPPLIER	ISIS LP.
INVOICE No.	89456
ORDER DATE	28-7-95
COPY No.	

630.11

Printed and bound by Hartnolls Ltd, Bodmin, Cornwall

To

Eva, Kenneth, Dion

& Julian Webb

and to

Vera, Norman, Elizabeth, Robert,

Gail & Dion Pritchard

this book is affectionately

inscribed

CONTENTS

1 All Holes & Corners 1
2 My Only Bid 11
3 Jessamy & Gold Rings 20
4 Cow Psychology 32
5 Perfect Summer 43
6 Nothing at all to Bill 61
7 Concerning Water 77
8 Gun-cat, Plough & Garden 85
9 Time Out of Mind 98
10 Hard Winter & Some Siamese Cats 106
11 Baths, Blossom & Plumbing 120
12 The Townsman in the Country 133
13 Calves & Bees 144
14 The Price of Solitude 153
15 New Gardening Policy 160
16 Harvest, Haying & a Treasure Hunt 168
17 Professional Gent 183
18 Burnings, Pigs & Skins 195
19 The Hut 203
20 The Hammock 212
21 The Fox Cubs 218
22 Silver Wedding 221
23 Milking Parlour & the Beetle 229
24 Punch Bowl Herd & a Transformation . . . 242
25 Summing Up 251

CHAPTER ONE

All Holes & Corners

You wouldn't marry a man two days after discovering his existence: especially without first having a good look over him and finding out a little about his past. You would probably also wish to ascertain, as far as was possible, what kind of life the marriage would entail for you. Broadly speaking, the principles of house purchase are similar. But, some might add, more so, because in these times it may prove easier to change spouses than to change houses.

In the purchase of Punch Bowl Farm we did none of these things. The place was practically unsought by either of us. One can only say that it thrust itself upon us. Admittedly we had played with the idea of moving some day, because we all longed for the real country and our semi-suburban house near Woking was the kind of house agent's dream generally described as, "Det. mod. house with all mod. cons., one min. shops and P.O., easy reach Woking, bus stop at gate." But it wasn't our dream: and besides, I had latterly begun to write books for children, and I am not one who finds it easy to evoke fields and forests when sitting near a busy bus stop.

Even so, it was a little surprising when Bill pushed the local paper across, one September Sunday afternoon, his finger planted on an agent's notice.

"Shall we go and look at it?"

I always read everything in the papers from the bottom upwards, so the first thing I saw was, "Sale on Tuesday next at the *Lion Hotel*, Guildford."

"There isn't time," I said. "The sale's in two days."

"It's a nice afternoon. We could look at it," said Bill.

"I suppose we could," I said doubtfully. "The children will be all right in the garden for a while."

These words were the first step on to the avalanche that was to transplant us, like a clump of alpine trees, from the known place to the unknown.

Reading on I discovered, while Bill got out the car, that Punch Bowl Farm was believed to be seventeenth century. It was constructed of mellowed stone and red brick under a roof of old tiles. There were massive oak timbers, concealed fireplaces and the original central chimney stack rising from ground-floor level. Standing in an entirely secluded position in country renowned for its scenic beauty and near the Devil's Punch Bowl, it was entirely suitable for conversion into a gentleman's country residence. There were seventy acres, thirty in hand and forty of woodland and heath.

"Aren't you ready yet?" Bill said in the doorway. "Ann and Shelley say they'll look after Shaun till we get back."

I said, "We don't want seventy acres, and it sounds rather ruined." But I was getting my coat.

The farm was fifteen miles away, and gradually the country grew more and more wild and remote. We were less than forty miles from Hyde Park Corner, but often for long stretches there was no house of any kind in sight; only pines and heather and bracken, unchanged perhaps for a thousand years except for the broad grey road running through it to the south. We found the village named on the sale notice and inquired for Punch Bowl Farm. The first man we stopped, though plainly not a "stranger here," had never heard of it. The next thought it over carefully for a while and then said would that be the old place way up the lane past the church, now? he believed it was for sale, but it was terrible out of the way, that it was.

We drove on, past old stone cottages with gardens over-brimming to the road's edge and into a slim green tunnel of a lane that climbed and twisted under its low roof, shutting out the view with its high hedge-topped banks. Except for the fact that we were still going upwards I quickly lost all sense of direction, so many and so sudden were the corners. It was, moreover, the kind of lane where you must hope you will not meet anything bigger than an Austin 7, because if you do, and have avoided hitting it, one of you will have to go backwards. Lanes like this always seem longer than they are, partly because of their beauty and partly because of their danger. One's attention is distracted between the two. This one was about a mile in length from the village till we reached a fork where a small signpost said, *Punch*

Bowl, Hindhead. Sandy Track Only to the left and, in large despondent letters, *Bridle Path, Churt, Headley. Impassable to Motorist* on the right. Neither way could be seen from the signpost, as the first one plunged at once into another of its dark capricious tunnels and the second dropped breathlessly out of sight and round the usual high-hedged corner.

We got out of the car and prospected. Surprisingly, round the corner at the bottom of the impassable-to-motorist piece we came suddenly on a large well-kept cottage with a double garage, a charming garden and at least four cats. The house was on one side of the lane and most of the large garden was on the other. Surely not . . .? No, indeed, how could we have thought it? One further glance ahead explained all. There was a wide farm gate across the lane. It was entirely arched over with spreading branches, and between the gate and the branches was framed the house we sought. One could hardly doubt it. There was the September sun on mellow stone and brick, there the roof of old tiles, sweeping in a cat-slide nearly to the ground. The roof had settled through its centuries into a lovely waviness, like the swell on a summer evening sea. From the gate, where we now leaned, we could see the crusted mosses on the tiles. Behind the old house and high above it a great black yew tree stood, full-skirted, sheltering. The house crouched down against it like a cat against a sun-warmed wall, looking across its leaning orchard trees towards the gate. Under the trees bowed with fruit the cows were walking leisurely in for milking, their slow hoofs crushing pears and apples in the grass.

"It doesn't look the sort of place that will be cheap," I said, stepping nimbly in my mind out of the ring of spells in which the house would have ensnared me. "Too large, too beautiful and too secluded to be a bargain. People will pay anything for charm and character, after all the austere years."

Bill opened the gate. "We've come to see it and we're going to see it. That is, if they will allow us. We haven't got an order to view."

Walking up the grassy track through the orchard we came to the farmyard, and in the middle of it, against a barn bowed with years, was a man whose head was in the bonnet of a lorry. We said good afternoon and was he the farmer? Taking his head out of the bonnet the man wiped his hands down a pair of oily overalls, the hands and the overalls remaining relatively as oily as before. No, he was not, he said, not so as you'd notice it; nor likely to be, neither, not in this set-up; he'd just been looking over it. Oh, we said, did he not think it a good farm? We had not seen it before. Good? he said; the house was in ruins, straight it was: and dark? you couldn't see across the kitchen, straight! And then there were practically no rooms in it, either.

No rooms? I inquired; but it was a *large* house, surely? I looked up at it, and it looked back through all its little windows, and I wouldn't be sure it didn't wink with the one boarded-up window in the attics under the gable.

Ah well, so it might be, said the oily one, but it was all holes and corners, see. There were only two bedrooms, what you could call bedrooms, and there wasn't any light, or water, or drains, and the what's-it was down at the bottom of the garden, straight! And the land! that was rotten, wouldn't grow a thing; the buildings were all falling down and wouldn't last a week, straight they wouldn't. And he put his head back inside the bonnet.

We walked on, wading through poultry, up the yard.

"Only trying to put us off," said Bill knowingly, "wants the place for himself."

We turned in to a garden drowned under nettles, picking a careful way over the stone path past a lovely tiled Well House to the open back door of the farm. At once it was plain that at least one remark of the oily one had been no overstatement. We could not see across the cavern of the kitchen. Bill knocked and the farmer's wife materialized out of the darkness.

Might we see over the property, please? if it was no trouble.

Had we, she countered, an order to view?

Well no, we hadn't, we were afraid. We had only seen the notice that afternoon, and being a Sunday the agent's offices would not be open. We were very sorry.

Properly speaking, said the farmer's wife, they were not obliged to show anyone over who had no order to view. They had had a lot of people, many of them just curious and with no intention of buying; and to-day was a Sunday, when they had hoped for a little peace . . . However, she said kindly — and how kindly, since they were the tenants, not the owners — perhaps a quick look over the house; she expected we wouldn't mind walking over the land by ourselves. We followed her into the gloomy kitchen, uncomfortably aware that we also were just curious, with so little intention of buying that one might say the intention was non-existent. Setting a brisk pace through the labyrinths she knew so well, the farmer's wife took us over the house. For my part I saw very little, because of having to use nearly all my powers of perception to see where I was going. But it was again very much as the oily one had said: a large dark kitchen, two large bedrooms and a great green sitting-room.

This little room? I inquired in passing.

Oh, they never used that, except for storing corn. The floor wasn't safe, and the door wouldn't shut, and the rain came in through the roof.

And this even smaller room, here?

Well, they didn't really use that, except for dumping things in. It was too small and too dark and you couldn't see out of the window without cricking your neck.

The cellar? The large well-lighted cellar?

Oh no, no one ever used that, it was a terrible place. Sometimes folk had clamped roots down there, as you could see by all the earth on the floor; but you should be careful how you went down, if you had to go down; the steps were all worn into hollows and covered with mud, and you were sure to stun yourself on the beam that went above them.

This little room on the staircase?

Well no, you see the floor wasn't safe . . . etc. . . .

Then what about this one, opening off here? And all that north gable?

But no one ever used that! There was only a rotting old priest-hole there and all the empty chimney-spaces belonging to the old wing which had fallen down time out of mind ago. You couldn't use that.

The attics, then? what about the attics? We had seen the boarded window from outside.

Oh well, you couldn't possibly use them; well, could you? with no floors, no walls, no ceilings, and the staircase falling into pieces. It stood to reason. Though goodness knew there was enough room up there, and folk had plainly used them at one time, long before any of us or even our parents had been thought of, she shouldn't wonder. And now, if we would excuse her . . . she was very busy. . . .

But back in the kitchen, pitying us I dare say, she kindly told us more. There was one thing she would say for the old house, and that was that it was dry. And all those great smoky fireplaces had been properly bricked up and replaced with nice modern ones. Admittedly the water was all to be wound up by hand from a hundred-foot well, but you got used to that in time, as you did to anything really. The place was lonely, as you might say, but then you didn't have neighbours to bother you. Haunted? Well, all she had ever heard was the wind in the big old chimneys, and the timbers creaking in their joints and that. You got used to it in time. And now, if we didn't mind . . .

Thanking her warmly for what was after all purest kindness on her part, we set out over the land. But if you have no map, and the boundaries are as self-effacing as they were on Punch Bowl Farm, you hardly know which is the land. It was easy enough with the thirty acres "in hand," for that was hedged and cultivated, but the forty acres of woodland and heath defeated us. Did it include the wild valley where trees in torrents ran down to a little silver brook far below? Or this high corner-piece — you could hardly call it a field — where purple heather stood above our knees and wild deer leaped before us into thicker shadows? Or this field of waving bracken from where three counties spread themselves below us like a carpet?

"Never let yourself be taken in by the beauty of a view," said Bill. "You have to think about the land."

We began to walk back to the thirty acres in hand.

"We'd have to let it," I said. "If we bought it, I mean, which we aren't likely to do. We only want about six acres, for the ponies and hens; and neither of us has time or knowledge to farm the rest of it."

"It's very poor land," Bill persisted. "Hot dry sand." He bent to grasp a handful and poured it through his fingers. "You know, I think that oily man was right. He may have exaggerated a bit, here and there, but on the whole he was right. Look. Couch grass, spurry, ragwort, bracken, gorse, thistles . . . and all that on the arable. However," he added, "what with that and the state of the house it just might go cheaply. You never know."

I said, "But it's so beautiful. I don't think I've ever seen a more beautiful place. Somebody very rich will buy it, and spend thousands on it, and make it a model farm with tastefully modernized farmhouse."

"Beauty isn't everything," said Bill. "I suppose you'll go to the sale? I shan't be able to, of course."

"I might," I said dubiously, "if I have time, and if nothing more important crops up. But it won't be any good. And after all, do we really want it? All that land — and the house needing such a lot done to it."

But of course, I went. It was asking for trouble, having had only the briefest glimpse of the farmhouse, little more of the buildings, and with no idea at all of where the boundaries were. Monday was an impossible day for further viewing, and the sale was on the Tuesday. On Tuesday I decided to go shopping in Guildford, and if time permitted I might look in at the sale.

Time permitted.

CHAPTER TWO

My Only Bid

In the sale room, arriving early, I acquired a catalogue. Now, for the first time, I held a map of Punch Bowl Farm and saw for myself where the boundaries were. The high heather field where the deer had flashed into the shadows was included; Punch Bowl field was its name and it was the farthest south-west outpost of the farm. The bracken field was there too; it had no name, which showed great delicacy of feeling in the farmers since the field adjoining bore the name of Little Bottoms. There was Hunters' field, where the yellow of gorse and broom would burn all summer; Hanger field, plunging giddily down to Rocky Lane and the wild valley; Barn field, like a valley in miniature itself; Yew Tree field, next to the farmhouse but higher than the roof of it; Upper and Lower Naps; Upper and Lower Six Acres; the Old Orchard; there they all were, coloured pink on the plan.

Opening the catalogue I glanced at paragraphs about taxes, rates, timber, shooting, and available services; about the character and charm of the farmhouse, its oak mullions and sills, leaded windows and oak-plank floors — no hint that one might fall through them — and about the old wing which was "pulled down" many years ago, so the catalogue cautiously put it.

I was reading when the bidding began. No, not the catalogue but, strange as it may seem, a children's fantasy called *Crowns* which interested me extremely. This may seem rather silly, but it shows how little I expected the proceedings to concern me. Besides, I have for years been an odd-minute reader, having too little time for all the wealth of books there is in the world, and have read while pram-pushing (on quiet country paths), while walking, beating cakes and batters, answering the phone, and even while bicycling.

The bidding started very low, distracting my attention for a minute, but there were several people in it and the figure crept up slowly. I returned to my book, still keeping a little awareness for the sale, but there was obviously, I thought, time to spare. It would be pointless to be one of several bidders, or even to be a bidder at all unless everyone else stopped. Suddenly everyone had stopped, and that before I thought things had properly got going. They'll all start again in a minute, I thought, and then realized with a great shock that the hammer was literally coming right down, and at a figure well below that which we had agreed as being our limit. I said two words — my only bid — while the hammer was dropping, and suddenly the sale was over. I had bought a ruined farmhouse and seventy acres that I had not heard of only two days before, that I hadn't seen properly and was not sure that we wanted. Some people are born with farms, some achieve farms, and some have farms thrust upon them. It seemed we were among the last.

Somehow I didn't believe it, in the way one doesn't when such extraordinary things happen. I stared vaguely and amiably in front of me, still holding my book open on my knee, while the crowd moved slowly from the room. The auctioneer beckoned from his desk and I went up to him. Name and address . . . sign here . . . (I was signing away the whole of our past lives, plunging my family at one stroke into the unknown) and now a cheque for the deposit, if I pleased.

"But I haven't brought my cheque book."

"Not . . .? Well, but . . ." No one attended a property sale, his voice implied, without bringing the elementary luggage. "Perhaps we can find you a blank cheque, madam."

"I hadn't really meant to bid," I said humbly. "It was just that I couldn't bear for it to go so cheaply."

"Well, it mightn't have, if you had let things take their normal course. Sit there reading through the whole proceedings then suddenly jump in like that with a two hundred and fifty pound bid when everyone else was bidding in hundreds. Crabbed the sale, I shouldn't wonder. Had 'em all thinking you were starting in to go on for ever."

"Do you mean the hammer might not have gone down when I thought it would?"

"Needn't have, needn't have. Sometimes we have to speed up sticky bidding."

"Well, I'm very sorry," I said, "if I really crabbed

your sale. But I didn't mean to, and I don't really think I could have."

He may have been right, for within a week we had received two offers of a thousand pounds profit from bidders at that sale. We turned them both down. If the farm was worth that much to others it ought to be worth as much, at least, to us.

Bill accepted the news of the purchase in his characteristic calm way. "If it's all right for you, it's all right for me."

The children, of course, were delighted.

"Have you come home with the farm in your pocket?" Shelley asked, looking up as I came in from the sale.

"I'm afraid I have," I said weakly. "Here is the catalogue, and the farm is coloured pink on the plan."

"Oh, *Mother*!" They were ecstatic, whirling me round and round the room; eleven-year-old Shelley, Ann her friend, and four-year-old Shaun.

"The house is in ruins," I said firmly, gasping in the middle of their whirlwind. "There's no water, no light, no drains, no bathroom, and the Other Place is at the bottom of the garden."

"Oh, *Mother*! As if anyone cared about that!"

When one is young one does not care about things like that; but when one is older they can matter very much. It wasn't the lack of modern amenities that worried me about Punch Bowl Farm, but there came a day when I was so worried about its state of repair that I got as far as writing to one of the two prospective purchasers. This was the Saturday after the tenants had moved, leaving the house empty, and we were able to make our first thorough examination indoors. The frightening thing was dry rot. Wherever Bill's probing knife was pushed, the rot was found; even my thumb could find it out in places. Bill, I admit, was not very much dismayed, but I was overcome with sudden dark doubts and fears. Could we ever make a home of such decay and ruin? Would the whole house collapse upon us all, one day? Would the great span of roof become too heavy for the weakening old rafters? Vacillation gripped me. And the worst of it was that Bill would offer no advice. I was to decide for myself, he said, since I was the one who would be living there the most, he being often away for long periods at a stretch. Adding that he saw no reason to alter his first opinion — that it was all right for him if it was all right for me — he left me to battle the thing out with myself.

I was very torn. The place was so lovely. The children would be happy there. We could have a cow for the house, and the ponies need no longer depend on hired grazing and bought hay. There were glorious ways for riding — no more tarmac roads — there was seclusion, and quietness for writing. There was the perfect ready-made background for my children's novels. On the other hand, there were ruin and inconvenience; all that land; the large garden buried under nettles; the water so far down and so laborious to wind up.

The two good offers tempted me. I think I vacillated for about a week, and then suddenly I knew we would go on with it. I said so to Bill, and he said very well, we'd better look out some old overalls and make a start. We started.

Once clear in my mind about the farm, all of life became exciting, assured and delightful. The worry was finished. We put our own house on to the market and began to spend all our spare time working on the farmhouse, driving down at week-ends with the children, Shelley's retriever pup, Glen, the tools and quantities of food. Mostly it was only Bill and I who really worked in the house, scraping and cleaning, cementing and distempering, but Ann and Shelley would put in erratic half-hours at beam-scraping, shovelling mud from the cellar, or painting doors and windows, before the outdoor regions called them and they were gone to the stables, the far fields or the valley. Shaun, at four, was more of a hazard than a help. He had, since his earliest days, been passionately interested in snails, and Punch Bowl Farm was a paradise for the serious snail fancier. We found them everywhere; among the tools, in pockets, on doorposts, and even in the lunch basket against times of scarcity at home.

The Old Orchard was littered with luscious fallen pears. Everyone ate them all the time, even Glen who consequently had colic in the car; but such was the pear harvest that autumn that the fruit lay rotting in the long grass under the trees: wings of red admirals, scarlet, velvet-black and white, crowded over them, for these butterflies rate rotten pears high among foods.

We took turns at winding water up from the well, and were amazed and a little discouraged about the time it takes to wind a bucket up a one hundred and seven foot shaft. But the water! was there ever such a delicious, sparkling, earth-cold drink? Ridiculous to suggest that it could be the same substance as that flat, lukewarm, chlorinated liquid familiar at home.

The slow wheels of the law ground round interminably while we fretted to move in and be on the scene of our labours all the while; but there were legal hitches about an obscure clause in the deeds, and we were negotiating two houses at once. It was some weeks after the purchase that we finally became the legal owners. There was still a lifetime's work to be done on the place, but it was now distempered throughout, most of the dusty old whitewash having been carefully scrubbed off (because distemper will not stay on over whitewash), and the mess of our work had been cleaned away.

We made the uprooting in November and moved furniture, ponies, dog and fowls in one glorious day-long muddle. Shelley's black pony Tarquin would not go into the horsebox, attracting the inevitable crowd of advisers at the gate. The dog had fits, the fowls escaped, the wash-tub with a wireless set inside it fell off the lorry and vanished for ever and ever. Half the furniture wouldn't go up the crazy farm stairs and the other half wouldn't stand level anywhere without two bricks, a tile and a couple of wooden wedges. We fixed beds with the help of a spirit level whilst lying on our stomachs, and hung pictures that swung drunkenly outwards from picturesquely leaning walls. We fell up and down the wavy cellar steps with things we couldn't fit in anywhere else and we made gallons of tea on a smoky black range that stuck truculently out in front of the banished and bricked-up inglenook fireplace.

We should probably never have had any lunch at

all but for the kindness of our nearest neighbour, Mrs Fisher, who, being a mere field away, was considered very near indeed, and who invited all of us to eat with her at her house, Rock Cottage: who indeed went much further than this, with offers of beds and baths and anything else she could think of. Bathing at Rock Cottage, I may say, was for over two years almost our only civilized gesture until our own endlessly discussed bathroom became a reality.

So we settled in and began our first long winter. We had not yet attempted any major improvements, but the house was fresh and clean, and most of the fine old timbers were scraped clear of plaster and rot, revealing to my unspeakable relief sound hearts beneath all.

We could now see across the kitchen without the slightest difficulty as a result of five coats of cream distemper, though the new south-facing window was not to be put in till the summer and all the noble old fireplaces were still hidden behind the usual modern monstrosities.

Bill was now in Wales for a fortnight at a time, so during that winter the planning and doing mostly fell to me. For the children all was delightful, of course. Not for them the worries about education in so remote a situation, the negotiations with water, electricity and telephone companies, or with farmers and agricultural officers who did not want to rent our surplus acres.

On schools alone our project nearly foundered. There were no early buses on the main road in the direction of the schools concerned, and Bill had the car. We did not want the expense of a boarding school for Shelley, and

Shaun would need a day school for several years yet. At last Mrs Fisher solved the problem by finding first an excellent governess for Shelley till the end of the autumn term, and then a satisfactory boarding school near Haslemere which accepted her as a day pupil. This school, about three miles away across wild country, had a stable and a paddock, so the transport problem also was solved. Every morning Shelley rode away on Tarquin; up over the farm fields, jumping the barways; into the rugged heath and pine land of Highcomb Bottom; through the great jungle of the Devil's Punch Bowl and over the notorious Gibbet Hill and Hindhead Common. What twelve-year-old could have wished for a better way to go to school, if go she must?

CHAPTER THREE

Jessamy & Gold Rings

Shaun, at home till he would be five in the summer, now went in for snails in a big way, and I began to fear serious snail trouble when the day came that I had time to tackle the garden.

"Look at the dear little souls, sliming along!" he would say, holding out his arms to assist their ascent of his pullover sleeve; and I would sigh, thinking of all the washing, and the water over a hundred feet down in the well, and no help whatever in the house because we were too far from the village. But the days of youth are short and should not be blighted by considering the washing.

At my time of life, on the contrary, one is often compelled to consider the washing: and when you bear in mind where the water was, and how busy I was — with the house and the children and the writing — you may rightly reckon it as madness that we bought a heifer and her calf. I should say it was really I who bought them, for Bill maintained his usual attitude, "I'm not here very much, after all, and it's you that'll have to look after them. If you think you have time, of course. . . ."

In fact I didn't have time, but one of the odd things

about life that I have frequently observed is that anyone can always find time for what they really want to do most. Some years later, when I was doing much more and yet still finding time for things I wanted to do, I remembered with mild astonishment my doubts about managing the cow.

She was called Jessamy and she purported to be a T.T. Guernsey. T.T. she undoubtedly was, for we had the vet's certificate, but Guernsey — well, she may have had a smattering; but we were very green in those days, far greener than we are now; though oddly enough I think we feel greener now, having reached that curious stage in the acquisition of knowledge and experience when one is suddenly overwhelmed by the idea of all there is yet to learn.

The calf, a heifer, we called Gold Rings, because she was born on the fifth day of Christmas, and she of course had first claim on her mother's milk. The dealer who sold us the pair had left instructions about feeding them, but no one can give you instructions on how to milk, any more than you can tell a person how to swim. These things come by practice only. I had once dabbled a little when a schoolgirl, and seemed to recall getting milk into the bucket, but a rehearsal was indicated before the cow actually arrived. A little way beyond Rock Cottage (which was the house we first saw, near our gate) was the next farm, Upper Highfield, run by a woman farmer, Nan Wilton, who was all that we were not. Her farm training and her own good sense together fitted her to manage the large herd of pedigree attested Guernseys, including a bull, and her sizeable mixed farm with efficiency and confidence, as she did. We were on leaning-over-gates-and-talking terms, so it was to Nan I went a week or so before Jessamy's arrival.

Yes, of course, said Nan, I could practise on old Lily; she was nearly dry now and easy to milk, so not likely to give me much trouble or I to give her mastitis. The thought of giving Lily mastitis (severe inflammation of the teats and/or udder) by my clumsy and unskilled fumbling almost destroyed what little confidence in the matter that I had, and my delicate squeezing at first produced nothing but a small peevish dribble. But luckily for Jessamy (and Lily) things quickly improved, and soon I found that my worst and only enemy in milking was weakness in the hands, wrists and forearms. There is only one thing that anybody can be told about learning to milk — except for commonsense details such as gentleness, scrupulous cleanliness and perfect calmness — and that is that if you close the top of the teat and then squeeze, the milk must come out at the other end. The rest is all practice, and more practice.

The proper time to take to milk a cow, said Nan, is seven to ten minutes. Cows vary, of course, some having smaller teats or narrower teat-holes than others and thus taking longer. A small teat is unbelievably fiddly to grasp, unless you have tried it. Nowadays I would not buy a cow, no matter how good, unless the teats were of a size to fit the hand or I had a milking machine in the cowshed. I have tried too much of finger-and-thumb milking ever to attempt it again.

Lily's teats were all that one could wish for, but still my aching muscles toiled at them, and the time I took to get half a gallon was longer than the time Nan took to fill a bucket. I would stretch my painful arm, flexing the stiff fingers, then bend myself to it again; I in a hot sweat with concentrated effort, though it was midwinter; Lily sighing with boredom as the minutes passed.

No matter what skill one is practising, or how ineptly, one improves in time. When Jessamy and her calf arrived in their horsebox I was still admittedly a novice, but no longer an absolute novice. I could milk. What I could not do was stay the course. In the milking line I was purely a sprinter, not a long-distance performer.

The first thing I did with my new knowledge was to take a sharp look at Jessamy's teats. For a moment I felt rather like an old hand in the market, looking over the beasts with knowledgeable eye; but pretty soon I felt like someone who has been a little diddled, because no angle of viewpoint could make the heifer's teats look anything but small. It was too late now; she had been paid for (excessively, of course) and was installed, and the horsebox was already rumbling down the track and under the arched trees of the gateway.

There was a sharp, deep thrill, I admit it, in seeing our first cow in the cowshed, our first calf, moon-eyed, lying in the yellow straw of the loose-box. Those rusty neck-chains, so long unused and bundled in the mangers — to fasten one round the warm soft neck of our own cow, to pour the ration of crushed oats and dairy nuts down into the low manger where hens had lately nested undisturbed; this was the real stuff of life. One could lean on the worn oak partition all the day, looking at her comfortable white-splashed form, listening to the clink of her chain and the slow contented crunching and snuffling over her food. But there was no profit in doing anything so trivial, for anything like so long a time. Shelley would soon be coming home from school, riding down in the dusk from the high fields and wanting her tea; and as to Shaun, there was no knowing, but with him it was always as well to find out.

Becoming gradually resigned to finger-and-thumb milking, I found I had somehow fitted Jessamy and her child into the crowded time-table of my day. It was largely a matter of getting up earlier, I discovered, and finishing the morning milking before the children's breakfast-time. And what breakfasts were these we had now! Coming straight from post-war milk rationing to the unlimited riches of a personal cow, we felt like pit-ponies turned on to clover. Now we were enjoying the full reward of our labours, the table graced with such plates of home-made butter, such bowls of crinkled cream and jugs of rich milk — as different from the bottled kind as our well water was from the piped supply — that we did not miss the fish, the sausages, the out-of-season fruits and vegetables that we had once thought indispensable. At last we had some small return to offer to our neighbour Mrs Fisher for her uncounted and continued kindnesses. Jessamy's bounty was so full that, even when Gold Rings had had her daily gallon of new milk, there was more than enough left for all our needs — though not nearly enough, of course, to bring the milk lorry up to the farm.

I made cheese and Devonshire cream, trying all the once impossible recipes having cream as an ingredient. Oddly enough, we did not at once grow portly, spotty or bilious, neither did we become tired of all this dairy produce; perhaps because we did, after all, use ordinary discretion in enjoying it, and so enjoyed it to the limit.

Through these first weeks of being cow-owners we were again much indebted to Nan and to her sister Jill. A cow's milk supply, Nan explained, would gradually diminish unless she were milked right out every day; and although I was getting plenty of milk (at some cost in sweat and aching muscles) she was afraid I was probably not skilled enough yet to strip the udder. So accordingly, every afternoon at milking time until she knew I was fit to leave in charge, Nan would walk along from Upper Highfield to milk the cow: or sometimes it would be Jill, but always it was one of them; and I would sit on the manger, scratching Jessamy behind the horns and marvelling at their speed and skill, at the great foaming froth in the bucket and at their ability to chatter about ballet, books, gardens, or cooking when I, in their place, would have been sunk in heavy concentration on the milking.

But all things are possible to those who persevere. Before February was far advanced I had discovered that milking times were the best of all times in a busy day for thinking out plots. I would sit, with my rubber-booted feet pushed into Jessamy's deep straw against the bitterness of the winter morning, going over chapter nine, perhaps, and hoping my cold fingers were not striking chill to Jessamy's inside.

There would be the usual two robins disputing for my bonus of crumbs upon the lid of the old churn; and all the quiet comfortable noises of the cowshed, with hungry overtones of urgency from Gold Rings, hearing and smelling her breakfast from the loose-box. I was well content.

But why, oh why, I thought, had no one told me how much water a dairy cow can drink? My original estimate, based on what the ponies drank and a little over, was ludicrously short. My day went something like this: rise at six, wind up water for cow, milk cow and feed calf; children's breakfast; wind up water for ponies, hens, house and cow; housework, evacuate snails, wind up water for cow; muck out cowshed; cook; wind up water for cow; lunch and washing up; take Shaun out on his pony; wind up water for cow and house; milk, feed calf; tea, with Shelley home from school; writing and typing; wind up water for anyone that wants it, particularly cow.

It is easy to see how much water-winding affected our first winter at the farm, and how high Jessamy ranked as a water consumer. We had considered harnessing the well with a petrol engine, but were afraid it might not stand the probable ultimate strains of bath, w.c., field standpipes and cowshed supply. So we waited for the main water, still no nearer than a pile of forms filled in triplicate. We had a further six weeks' frustrating delay because a form we had to fill in was out of print. One would have thought that someone could have typed a copy for us, but there is no imagination in bureaucratic circles.

During the winter we had tried to let our surplus land, but without success. Nan's herd was attested and therefore not allowed to cross the road, and she was not short of arable. Another farmer had a spade-lugged tractor, unsuitable for driving on a road, and another was too far away. We tried the Surrey Agricultural Executive Committee, but they were now giving up the land acquired in wartime. Slowly the coils that turned us into farmers were beginning to wind around us.

Bill was still away a great deal, and though my mother was now living with us (relieving me of such a burden indoors that I began to know again the sheer joy of just standing and staring), I knew that I alone could not farm the thirty acres now lying idle. We had no implements, even had I the time and knowledge. That winter we compromised while marshalling our forces. We lent two fields to a farmer in the village and in them he grew wheat. A third, smaller field, we had ploughed and sowed for us by contract with dredge corn for the cows, ponies and hens. The rest we used for grazing. But we knew our time was approaching and we viewed the matter with mixed feelings. Neither of us wanted to lay out capital on implements, to spend long hours of time on the land at the expense of other more profitable work, to risk the heavy losses always lying in wait for the novice farmer.

There was, however, far too much to do each day for me to spend much time in worrying. In fact, I began to give up worry altogether, and found I didn't miss it. One had always read and heard, of course, of the virtue of living for the moment, but why do most of us take so long in discovering this for ourselves?

"Because I always think there might be something I could *do* about it, if I go on worrying," a friend once said to me.

Thinking this over, had anything I had ever worried about, I asked myself, been much improved by my worrying? And the fullness of life that awaits the ex-worrier has to be known to be believed.

The last worry I gave up myself was my worst one, and it was a by-product of living at Punch Bowl Farm. Fairly soon, I knew, Bill would have to give up his job in favour of farming, or we would have to leave; because, quite rightly, no one is allowed to hold farm land who doesn't see that it is farmed, and farmed properly. It would be a long time before the farm could pay, no matter how little, because of our small available capital and smaller experience; thus, for an indeterminate time the whole establishment would have to depend on my earnings as a writer. What, I used to think in the middle of the night, would happen if one day, suddenly, I couldn't think of a plot for a new book? This was my favourite worry, and it was as I say the last one I relinquished. But it had to go. Worrying is so bad for one's writing. And, six years later, we are still at Punch Bowl Farm and I am still writing.

The first job I ever did on the land was to pull turnips for the cows and ponies. These turnips were the only crop left unharvested by the previous tenant and they were in a little hilly field that we planned to plant as a mixed domestic orchard. It was great fun working out on paper the sixty different varieties of apples, pears, plums, cherries and peaches, all the best in their kind, that would be planted during the next winter; but meanwhile one had to get on with the daily turnip-pulling, whether it rained or not.

I well remember one such day, with the rain pouring down the back of my neck as I bent above the rows, pulling the roots and throwing them over the fence into the pasture. The dog Glen was with me, but I had no eyes for him, keeping bent into the rain and my work. Now that dog knew he was a retriever, but in his muddled mind things got a bit confused. After about half an hour of pulling and throwing I straightened up and looked to see if I had done enough. There was not a turnip in the pasture, but all were safely retrieved and laid neatly back upon the turnip rows. I never see a turnip but it calls that dog to mind.

We made a room for my mother in one of the holes-and-corners; the one that used to be a granary ("Still is," said Shaun, "with Gran inside it"). This room, like most of the house, had been whitewashed throughout. As any decorator will confirm, you cannot distemper over whitewash, for, if you do, the distemper is certain to flake off. There seems to be no labour-saving way of removing whitewash. Either you brush it off with a stiff wire brush (in which case clouds of it fill your hair and mouth and nostrils and seep through into the house), or you wash it off, which requires quantities of water and is also very laborious. I washed it off, and for this purpose I carefully saved all moderately clean rinsing water for several days together, fresh water being far too precious to use for scrubbing and cleaning. We had scraped and brushed the oak beams before we moved in, but the floor was in need of repair. Bill cut out the worn and rotting patches and fitted in new wood; and then he strengthened the whole by cutting upright joists to measure and fitting them under the floor beams, in the outer cellar directly underneath. (We still had to tackle draughts, later on, when we found that mother's carpet used to take on a steep swell in a wind — she said it made her feel seasick — and this Bill and Shelley did by plastering brown paper cement-bags under the floor, across the cracks.)

The walls were distempered pale yellow (a pleasant, cheerful colour, despite the modern tendency to call it "margarine"), the door and window-frames were painted to match and the room was ready for mother.

This entirely successful venture encouraged us to make a little room for Shaun out of a second hole-and-corner; the one where you cricked your neck in looking through the tiny high window. The first essential in this room was a new window, larger and lower than the old one. We ordered the window — leaded in the old style — and Bill began to cut the wall to fit it. This was no easy job, the wall being eighteen inches thick and made of irregular hunks of Bargate stone. Hours of work with a hammer and chisel were required before the window space was brought to size and literally sculpted to a perfect rectangle. The window was finally fitted, puttied and painted, and the deep ledge tiled with red quarries.

The floor in this room was in fair condition, requiring only slight repairs, and soon I was saving the rinsing water again and scrubbing off whitewash. Mother was here now, and helped with the distempering (we took the same butter colour right over the ceiling, to make the little room look larger), and it was she who made the linen curtains for the new window.

The finished room was just large enough to take Shaun's divan bed, chest of drawers, bedside table and stool. We curtained off an alcove for a wardrobe and arranged his books across the top of the chest of drawers. His window, opening to the morning sun, looks out on the farmyard, the drive, and the Little Orchard, then still the turnip field.

And now the house had four good bedrooms where it once had only two.

CHAPTER FOUR

Cow Psychology

The springtime of that year was the loveliest I remember, starting in March with warm still days of sunshine, summery enough for outdoor meals and clout-casting. Gold Rings was allowed the freedom of the farmyard in the daytime, and she would amuse herself by engaging in a personal steeplechase, round and round the yard at great speed, jumping everything that lay in her path. But if Jessamy should be grazing within sight in Barn field or the Old Orchard, Rings would start a sustained bawling at the gate. Jessamy was not very motherly, but she was the type that cannot stand a child's crying. Had she been human she would either have picked up her baby and jiggled it, or thrown it out of the window. Being bovine she thought along different lines. Hurrying to the gate, she would open it by poking her horn under the catch, thus letting her child into the field. Rings would dash through at once, doing the whole afternoon milking for me and overstraining her own four stomachs, all in five minutes. No human can ever milk as fast as a calf can, even though a calf cannot milk two teats at once, as we do. Jessamy herself would take no further interest in her child, continuing her grazing with the most profound indifference, until Shelley or Shaun or I would see the state of affairs and rush to fetch Gold Rings back.

About this time I was asking Nan's advice again, and again it was about Jessamy who had lately developed alarming symptoms of delinquency; she, the gentlest and most resigned of cattle, had become a psychological problem. It began one mild April morning when I walked beneath an ecstasy of nightingales, down the grassy track past the children's great Swing Tree, to fetch the daily paper from the gate. Jessamy had been mooning as usual by her swiftly emptying bucket at the farmyard gate and, bidding her an amiable greeting, I swung on. Before I had reached the Swing Tree there was a sound of galloping hoofs behind me.

"Ponies have got out," I thought, glancing backwards, and there was Jessamy, she of the sad eyes and meek, demure step, coming fast towards me with her comical white head lowered and her tail stuck straight upwards like a mizzen-mast.

It was nice to think afterwards that, though approaching middle age, I had yet gained the Swing Tree before her, despite her advantage in youth and extra legs. There I grasped a stick and beat off her repeated sorties until she decided I was still master and not to be trifled with. From then onwards she recognized a stick, and with one we could venture safely wherever she might be.

"Her trouble," said Nan, listening to my tale, "is most likely that she is all alone. Cows are herd animals and they get soured if you keep them alone. Have you thought of getting another one? It would probably put a stop to all this."

So in came Dawn and Duchess.

Duchess's full name was Norney Countess's Duchess and it was no misnomer, for this cow — far from being the rather vulgar type that Jessamy and Dawn were, belching loudly and happily in the cowshed and walking like charwomen — lived up to her title and was a most dignified and gracious matron, being much better bred than myself and not at all the kind of cow whom you could scratch behind the horns and slap upon the rump. She kept her distance, she expected others to do the same. I always felt that I should curtsey before milking her and walk backwards out of the cowshed, though as she came to know us she unbent enough to permit a little discreet familiarity.

Whether she and Dawn would really have cured Jessamy's mental disorder we never knew, because, feeling that two cows were quite enough with the present water situation and bearing in mind the finger-and-thumb milking, we sold Jessamy and her calf in part-exchange for Duchess, which left us with what we had really wanted in the beginning — two pure-bred Jersey cows, the one in calf and the other in milk.

Among her other charming aspects Duchess kept a mouse. It was a brave and stout-hearted little mouse which at first I took to be excessively in the plural. Milking Duchess on any morning of that rare and exquisite spring I would be aware of much traffic in her manger, much coming and going and squeaking and excitement. Sometimes I would see a tail before it vanished under piled hay, or sometimes a whole and enchanting brown mouse, so small, so soft, sitting upright in the manger, its tiny hands clasped round a single oat-grain, holding it up to nibble as a squirrel does a nut. Six inches, perhaps, above its head would be Duchess's broad black muzzle, blowing, chewing, snuffling; and then suddenly there would be a deeper, louder snort and away would flash the mouse along the mangers, hay-stems bending under her small feet, and vanish out of my charmed sight.

For a long time this shuttle movement would go on, mice coming, mice going, and I thought the whole place must be infested until I began to realize that it was all the same mouse, and she was taking home the rations to her babies, one grain, one journey.

Until Duchess got used to my slow milking — I was never able to improve my time very much — she frightened me badly once or twice by going to sleep while I milked her and all but falling down on top of me. In time we became adjusted to each other, she to my slowness and I to knowing the first signs of her drowsiness. After about nine or ten minutes she would show her impatience with me by pressing her nostrils flat against the side of the manger and blowing through them with a curious, bored sighing noise. A minute or two later, if I were not watchful to stop her with a slap and a shouted "*Duchess*," she would drift off into a doze, and then suddenly begin to totter. When you are squatting on a low three-legged stool, with a bucket clamped between your knees, the sudden prospect of a ton of dairy cow about to fall upon you is fairly petrifying. But as I say, in time we grew to know and respect one another, and I have known her to come and lie beside me, chewing blissfully, when I have been writing in some corner of her pasture.

With Dawn you could do anything; she had no complexes. The children sat astride her, I leaned on her in contemplation or hung my milking overalls across her back. She would stretch out her neck like a kitten to be stroked underneath it, and then she would sigh blissfully, lowering her amazing long black lashes over her large violet eyes. She was yellow as a primrose path all over, including her muzzle and tail-switch which should by Jersey standards have been black. She came when she was called and was not ashamed of sentiment; had she been human she would have been reading Florence Barclay and having a good cry in the cinema.

That spring we had two cowshed kittens. The theory was that they would live there and have their bread-and-milk there, and hunt and grow up to be cowshed cats. In practice, as soon as they were old enough to follow me from the milking they were in the house and rousing such hostility in my Siamese Vashti that we found them good homes elsewhere.

But when they really were cowshed kittens they were delightful. They used to swing on the cows' tails, as children swing on ropes; a run, a grab, and away they went, hanging on Dan's or Duchess's wavy tail-switch. They stalked the hens, climbed the door-posts and hid in the nettles round the corner of the cowshed. At least, they hid in the nettles until one day when they accidentally got the full baptism of chlorinated udder-washing water which I always flung there hoping it would discourage the nettles. I had for the moment forgotten about the kittens, but they never forgot about that.

They used also to lie in dark tight circles in the yellow straw beds of the cows. Seen from the garden, through the open cowshed door, they looked just like little cowpats, and twice I have rushed down there saying to myself, "However could I have overlooked that cowpat when I was mucking out this morning?"

The kittens always had the foremilk from the cows, the first streams that may have dust in from the teats. As time went on it was difficult to stop them from climbing up the cows' legs for the milking of the foremilk, and they always jumped on to my shoulders as I squatted with their little dish beside the cow.

Sometimes, about this time, the children would come running down to the cowshed with their mugs in which they had measured blackcurrant syrup or sugar and fruit flavouring, holding them out for me to fill with new milk frothing straight from the cow. The result was a sort of super milk shake, tinted, fruit flavoured, and crowned with an inch or more of coloured foam. The only thing it lacked was icy coldness, but no milk shake could ever have been fresher.

Meeting Nan one day and talking over the cows as we used to, she said, "I suppose you'll be thinking of getting Duchess in calf again, soon. See, when did she calve last?"

"About a month ago," I said, calculating. "It was a bull calf so we didn't buy it with her."

"Then you'll have to wait another couple of months or so," said Nan. "They take nine months and you want to aim for a calf every year. Let her miss next time and then catch her when she's bulling the time after."

"Yes," I said, "but how do I know when?"

"Oh, you couldn't possibly mistake it," said Nan.

"I dare say I could," I said; "you forget what an amateur I am."

"You'll know," said Nan, adding, "I'm afraid my bull wouldn't be any use to you as he isn't the same breed. You'd better have A.I. . . .* You get the best bulls, and they come straight out the same day you phone them. Not the bulls, of course," she said, doubtless uncertain of just how amateur I was.

"No, of course not the bulls," I agreed. Really she mustn't think I was as silly as all that. But still — how did you tell when A.I.-Day had arrived?

Touching the subject while talking to Mrs Fisher one day, she said, "They make a lot of fuss, you know. You'll soon see," and then told me a nice little tale about a friend who had written to her that morning. Among the news was an item about the friend's house-cow, Ada. "Roberts," said the friend to her odd-man, "what are we going to do about Ada? Every time she comes back from the bull she gets more and more temperamental."

"Well, Mum," said Roberts, "I don't know I'm sure, unless we has her artificially incinerated."

(Just to save anyone the trouble of writing to tell me that this story isn't original; they saw it in *The Countryman*; I now explain that it was I who sent it there.)

*Artificial Insemination.

We had a good laugh about this, and I admired Mrs Fisher's lovely garden now dancing with spring flowers, and she inquired when I would be making a start with our jungly plot. Perhaps next year, I said, wishing there were more hours in a day and much more energy in the human frame.

Duchess, as I have said, was a very decorous cow: but bulling, I soon found out, is a most disorganizing event in the life even of the stateliest cow. Dignity is forgotten, food untouched, milk held back, and you do well to look to your fences. Bulling cows know to a point on the compass exactly where the nearest bull is to be found, and during these first years our nearest bull was Nan's. Duchess would tear from the cowshed after milking, straight to the top north-east corner of Barn field, this being the nearest point in her field to Nan's bull, and there would languish and bellow and moon all the day. This spot became known to us as Bulling Corner (I once made the great mistake of having my writing hut erected there. It didn't stay for long), until Nan married and went to India, and for a time there was no bull at Upper Highfield. Immediately our cows located the next in order of distance. This was a Jersey gentleman at Pitch Place, away across the wild valley in almost exactly the opposite direction, so that Bulling Corner was temporarily displaced from its normal location and set in the south-west.

When Duchess's wedding day really broke out we were taken by surprise. Bill and I had planned to attend a farm sale, with a view to our imminent farming future; Shelley had left for school on Tarquin, and there were only mother and Shaun to guard the farm. My mother is not the rural type. After half a lifetime spent in being an admirable wife-cum-curate to my clergyman father, she did not take with gusto to farm life in her retirement. She had always feared cows, and consequently they had always haunted her: in her dreams by chasing her, and in real life by fancying her company. There was, for example, the time when she was sitting in church one hot summer Sunday and, the door being open and the cows outside overheated and bothered by flies, what more natural but for a cow to come into the church? Father saw the entry from the pulpit but, being a man who never did anything without much deliberation, the cow was sighing down my mother's neck before she heard it above the murmuring of voices.

"I had better stay at home from the sale," I said. "We can't possibly leave you to cope with Duchess, Mother."

Mother said don't be ridiculous, or words to that effect, she was living on a farm now, and was going to take her share of the farm's perils.

Before I had got over my amazement at these words Bill had thanked her very cordially, said what a real sport she was, he didn't think she had it in her, and so the whole thing was arranged out of my hands. I did my best to leave all in perfect order, fetching Duchess down (with difficulty and guile) from Bulling Corner and chaining her up in her stall with water and hay. I put out soap and a towel for the A.I. man and phoned the Cattle Breeding Centre. They would send a man out during the morning, they said. The charge would be one pound, covering two more visits if the first should not prove successful.

"All you have to do, Mother, is to give the man a bucket of hot water, pay him, and ask him for the ticket with the bull's name on it. You needn't go near the cowshed; we'll let Duchess out when we get back."

We had a wonderful day at the sale, buying a second-hand spade-lugged tractor for fifty-two pounds, an old horse-plough for — literally — a shilling, and various small items in harrows and other farm etceteras, thus throwing another spadeful to bury our past lives.

Driving home in our elderly Standard we felt quite agricultural when we noticed a wisp of straw stuck in the mudguard. Presently, we would have hay in our hair, swill in our boots and tractor-oil on all our clothing. I cannot say about Bill, but my pleasure in our return was a little damped by apprehension about mother and Duchess. But it was another example of the futility of worry. Moreover, I had underestimated Mother.

Hastening to the yard, I looked across to the cowshed. The door was propped open to the westering sun and a cheerful nuptial bouquet, tied with ribbons, was swinging from the lintel.

I raced into the house.

"Mother, you old fraud! Scared of cattle! You let her out, too."

"Well dear, she seemed very quiet after that nice man had gone, and I thought it was a pity to keep her in, in all this sunshine; and Dawn was calling for her, too. I undid her chain from the other side of the partition, and she just rushed straight to Bulling Corner."

"Well, honestly!" I said. Then turning to look out of the back door, "She's still there now!"

"Did you like my bouquet, dear?" asked Mother, putting the kettle forward.

CHAPTER FIVE

Perfect Summer

Our first summer had a dream-like quality in that it was neither the old life nor the new. We were not yet very deeply involved in farming, and there was more time merely to enjoy the place than there has been ever since. For one thing, my mother was still with us through that year, and so my indoor work was halved at least. We spent more time in planning for the farm than in working on it; but nevertheless we were much more busy than most non-farming families we knew. Bill still had his old job, and in his free time worked on the house or at hedging, ragwort pulling and the like. I was slowly advancing on a broad front into the nettles, old bricks and tiles, and other rubbish of the garden, besides doing my usual house and cowhouse jobs; and somehow I had to fit in the writing of two books a year, with articles and short stories.

In June we made hay for the ponies and the cows. It was a leisurely haymaking, from only an acre of mowing grass; unlike those of our later summers when eight or nine acres had to be cut and made and carried with only family labour — and, the haying being in June, and therefore in term-time, family labour was usually just Bill and myself. For our first haymaking we had no tractor or working horse, no mower or tedder or rake or proper cart. The acre was cut by scythe, mostly by Bill with some help from me.

Scything, like most manual arts, is extremely hard work when you are learning, and so easy when you have learnt that it is almost a pleasure. I was learning through nearly all of my scything, that June; trying to teach myself the effortless swing in which no muscles pull and no tendons strain; but, by the time the work had become a pleasure, there was little left to cut. What a delightful thing scything is, when the struggle has gone out of it! Imagine the hot June day, the sun warm on bare arms, the light wind stroking over the tall grasses as a hand strokes ruffled fur. There is pollen dusting on one's shoes, for the sun has soaked up the morning dew. The scythe is sharpened; its lovely shape hangs easily from the hands (all the old hand-made implements of toil were lovely in their shape; the plough, the wagon, the yoke, the axe, and many more). With a slow smooth swing of the shoulders one sets the scythe mowing, putting to the sword flower and grass and weed alike, the swathes falling rank on rank beside one as an army felled in battle; moon daisies, moth-like in their white airiness, small scabious bluer than the sky, bold yellow hawk-bit and rusty-brown sorrel, all giving up with the grasses their essential goodness to the haying.

The movement is hypnotic, the slow measured stride advances down the field, the steel blade slipping through the dry stems with a whispering hiss. And as the sun gains power the indescribable sweet scent of drying grass and clover fills the air.

Our farmhouse lies low under the hills, and all the fields around are higher than the house. When we make hay the scents of it fill the farmhouse, floating in at the open windows and haunting the rooms until the hay is all packed away inside the barn. The short warm nights are soaked with the essence of dew on wind-rows: what the nightingale was to the nights of May the drying hay is to the silent nights of June.

This, our first hay, cut with the scythe, was raked with hay-rakes made by Bill from wood he had felled on the farm. All four of us raked it, and the weather was the best that hay could have. There were upturned mugs beneath a haycock, with bottles of lime juice and orange squash. On, in and around the haycocks were the cats. How they managed to avoid being raked and cocked and stacked I never knew, for in the heat of those long days we grew too tired to be watchful.

The day came when, the hay being tested between the teeth, no moisture could be detected; it was fit for carrying and stacking. We carried it in Red Clover's little tub-cart, with extensions fitted to the front and sides. This method seemed exceedingly precarious, when moving, loaded, down the steep hill, with three of us hanging desperately on to the rear of the little cart; but all was carried safely and stacked inside the barn.

Now that we are accustomed to the ubiquitous hills of this farm, the strangeness is more in moving on a comparative level. We have no absolute level anywhere on the land. And with the tractor and trailer there is no need for hanging on behind; the weary workers all ride home on the load-top, burrowing as flat as they can into the heady harvest when the track passes under low-hanging branches of the hedges.

Still I remember, in that first summer, walking or riding just for sheer delight, and not with a gun or a thistle-spud or knapsack. The cats and Glen would always come too if not thwarted, but sometimes we would leave Glen at home, in the hope of seeing a fallow or roe deer, the badgers in the valley, or a fox.

To see the badgers one went on foot, for quietness and because of the precipitous nature of the valley, and one always wore special clothing chosen for its fly-repelling qualities; up at the neck, down at the wrists and ankles and with no gaps anywhere at all. The midges and mosquitoes were always populous in the valley because of the shadows and the stream; but at the badgers' awakening time — about an hour before dusk — the bloodsuckers dined. If you were the one they dined off you would never see a badger, because the first secret of successful badger-watching is complete stillness; the others of course being good camouflage (you don't wear a pink blouse and white skirt), silence, and a squatting place downwind from the setts.

There were few things more enchanting on a fine spring evening than to sit thus well-protected from insects, under a mist of young leaves, listening to the sundown concerto of the birds in which the nightingales so exquisitely played solo. And if the badgers should come out to charm you with their light-hearted family life — leaping and rolling with their cubs and squealing their pleasure in the evening — then your hour was made perfect.

Walking home through the high fields in the first dark hour, the sounds of night would be clearer and more dramatic than any sounds of day, for in the darkness one's ears are more important than one's eyes. A goods train running slowly through Haslemere, three or four miles distant, might be winding round the rim of the Punch Bowl, so clearly does the sound float over in the still air. Tawny owls in the dark ravine make their loud reiterated statement above the perpetual snoring of the nightjars. How the nightjar makes his drone I do not know, but I have heard him hold one long throbbing note for a full five minutes before it was cut off, as with a knife, to start again a minute or two later. Coward, in *Birds of the British Isles*, suggests that the "trill" as he calls it, is made in the throat, for he says: "the lower mandible vibrates, the throat is distended until the feathers stand out." There is nothing "jarring" about this soft, slurred, drowsy trill.

The nightjar, a late spring arrival to Britain, has other more expressive names: Jenny spinner, no doubt inspired by the spinning wheel whirr of its song; fern owl, perhaps because of its haunts; night hawk, on account of its food; and moth-owl, because of its silent moth-like flight. Other names, such as goatsucker, flying toad and lich (corpse) fowl, seem ill-applied to me. Certainly nightjars do not suckle goats, any more than hedgehogs suckle cows (and supposing they did, how odd, in view of milk capacity, is the country comment: "She's three pints down this morning; I'll lay there's been a hedgehog at her").

Now, in the night, the drama of the rabbit's warning *thud-thud* is only less tense than the heart-clutching sound of its scream from the jaws of a hunting fox.

Clouds floating from the moon reveal a black-and-silver landscape of tall wild hedges dark against the sky, and heather like a grey sea rolling down the land. Massed trees look like thrown-up volcanic rocks, single ones taking on strange shapes and personalities against the summer night.

White flowers show up like silver coins in the grasses, and white moths flutter in the face. The grass is wet with dew, making a lush sound against one's boots where in the daylight it had rustled dryly. The air is cold, it is good to see the light from the farmhouse kitchen flooding out through the open door and to hear the cheerful sound of whistling inside.

This was the summer when Shaun was five and began to attend the village school. I would take him down the long shadowy lane on fat Red Clover, riding her home myself when I had seen him in at the school door. Clover was and is one of the chronic not-so-slim. She is the kind that eats because she likes eating, and not merely to satisfy hunger. Probably she doesn't know what hunger feels like, though in the summer we usually try to shut her in the stable in the daytime so as to quell a little her vast intake.

In the afternoons I would take her down the lane again to bring Shaun home; one can hardly call it riding her, but I would sit on her broad back with my legs dangling well below Shaun's little stirrups, and she would patter along gaily, well up to my weight and willing with all her stoutness. She reminds me of a horse-chestnut, round and red and very shiny.

It was difficult, I found, to get away without the cats on these school journeys, especially without my Siamese Vashti. She is a passionate walker, but only if I am going too. She is a one-man cat by nature and had developed a "fixation" on me very shortly after her arrival, and this was how it happened. She *would* talk. Probably she did only talk when she had something to say, but she had a terrible lot to say. First there was telling us all about how awful it felt to be new, and how nothing in our house was done at all the way she was accustomed to. And as well as this there was a lot about what she thought of Shelley's tabby Patrick, and about the food and the English climate, and about having a dog in the house and so forth.

Now my mother was, as I have said, not very fond of animals, but least of all of cats. It was noble of her to support life with our two at all, but gradually I saw that Vashti's conversation was getting more than she could bear. I would have her neutered, I said, sinking my original plan for Siamese kittens; that ought to make a difference. She went to be neutered. Bill and Shelley fetched her home, but they had some shopping to do in Godalming and left her in her basket in the car. Sometimes a Siamese voice is so much like a human baby's that anyone can be deceived. Bill and Shelley found a woman peering in at the open window of the car.

"She thought we'd got a baby in the basket," Shelley said. "I told her it was a Siamese cat but you could see she didn't believe me, so I opened the lid just a little bit to show her, and I told her Vashti's name. She got quite friendly then, and would you believe it, Mother, but her name was Vashti, too."

Quite probably these were the only two creatures in the whole of Surrey bearing that unusual name, and there they were, peering at each other through the wicker of a basket.

My hopes about a cure for Vashti were sadly dashed. Now she had a whole new field for conversation: "Have I told you about my operation?" Far from being any better, things grew worse. We would give her a few more days to settle down before we judged her, I said, but a few more days went by with no improvement. I had defended Vashti staunchly all this time, being almost her only supporter; but now I said I would give in, I would write a letter to a friend in Sussex to whom I owed a half-share in a Siamese since an earlier sharing arrangement. I wrote it, and left it ready on my desk for posting, and I felt like a betrayer.

All the rest of that evening there was quietness. At bedtime Mother said, marvelling, "Do you know, I don't believe I've heard Vashti once this evening."

No, I said, neither had I; and in the morning I thought I would not give that letter to the postman; perhaps to-morrow. . . .

But all that day the quietness was profound. And the next day, and the next. After three weeks I prudently steamed off the stamp and then burned the fateful letter.

Vashti is now five years old and practically never speaks unless she is spoken to; but you can understand the one-woman fixation, can you not?

* * *

For the children this summer was a perfect one, long to be remembered and talked of. Shelley and I rode many miles on hazy summer evenings, all about the Devil's Punch Bowl and the windy heights of Gibbet Hill, she on Tarquin and I on my chestnut Nanti, who never walks if she can dance, never dances if she can canter and never canters if she is allowed to gallop.

More often Shelley rode with Ann, who paid us long visits in school holidays as she had done since she and Shelley were quite small. In those earlier days both Ann and Shelley from time to time gave me bad shocks by rushing in and saying, "Something awful's happened!" and then, after I had died several deaths for whichever of them was not visible, or for Shaun or Bill or both the ponies, they would go on with, "Tarquin's got out of the paddock," or "Clover's cast a shoe," or some such lesser note of bad news.

Realizing what a whole new field they would have on a farm for these dramatic but heart-straining announcements I said one day, "Please, both of you, don't ever, any more, start off any sentence with 'something awful's happened.' I can't tell you what it does to my inside. Even if it is something awful — which it rarely is — break it gently."

Forgetting all about this appeal, I was peacefully milking one afternoon a little later when Ann came strolling down the yard to the cowshed doorway.

"Hullo," I said, "have you enjoyed your ride?"

Oh, yes, said Ann, very slowly and quietly, they had enjoyed it very much: it was rather windy in the Punch

Bowl, of course, but nice and sunny; both the ponies were all right, they were in the orchard now; she was all right, fine in fact; Shelley was pretty well all right, too; not really hurt, you know; she was in the sitting-room wrapped up nicely in a blanket; I wasn't to worry; it was a pity Shelley had been sick on the kitchen floor, but Ann had dealt with that; and now Shelley was sort of asleep, or a little bit dopy, something like that——

I dumped my bucket in the spare partition and tore up to the house. Wrapped in a blanket, as Ann had so gently broken to me, Shelley was huddled unconscious in an armchair. Now that the shocking truth was out, Ann rushed on graphically into a full account.

They had taken the ponies out without saddles or bridles——

"But Ann," I said, "you both know you oughtn't to do that! you've absolutely no control without a bridle. And no riding hats, too!"

Yes, she supposed it was a bit silly but it had seemed fun at the time, like Indians, riding by hand-touch. Everything had been all right till they got on to the high rim of the Punch Bowl, and then the wind sort of went to the ponies' heads. Tarquin had started bucking first, then Clover. Both she and Shelley came off after a bit — you simply couldn't stay on — then the ponies galloped off. She herself was all right, as she fell in the heather, but old Shelley must have hit a stone; she was very queer walking home — doddery and feeling sick and that kind of thing — but she had insisted on catching Clover while Ann caught Tarquin. Then they had brought the ponies home (a full mile) and turned them out into the Old Orchard, and then Shelley had suddenly been sick, just like that, on the kitchen floor, and she said the room was rocking about and then she went all funny, and Ann had quite a job getting her into the sitting-room.

It was concussion, of course, and Shelley had to spend three days in bed.

For a long time my rigid rule was followed; riding hats and bridles *whenever* you go riding. But with Ann's and Shelley's kind of riding this rule was doomed to fizzle out. Both girls would catch a pony at any time and in any place, wherever a pony might be found, jump on and ride away anywhere on casual impulse without even a halter for control. They would, neither of them, dream of walking down to the gate if ponies could be caught and made to carry them; they vaulted up with full hay-nets, riding with them from the barn to the ponies' shelter; they would set out on foot to pick blackberries, see the ponies in Lower Naps and return on horseback.

I became immune to "something awful happening."

Shelley later set off with her cat and dog and pony to a nearby spot in the high fields and went into solitary camp. Sometimes a friend was invited to join the camp, but other days and nights she was alone except for her animals, and we saw little of her except when she made sudden raids upon the farm to fetch supplies, water, saddle-soap or a clean shirt. Glen, she said, was all the guard anyone could need. Sometimes he did rush from her tent in the dark hours, barking and growling, but she believed it was usually rabbits.

Shaun rarely plodded up the fields to bother her but took on stronger meat about the farm, for he was five and getting on; he learned to drive the tractor, becoming coolly expert and only betraying his extreme youth by christening her Starlight. He never lost his interest in snail life and in what he called hairy catalogues, however, and began not only to cultivate and cherish all our own but to import from neighbouring places; so that I continued with my garden against snail and catalogue trouble as well as the usual Siamese cat trouble, cow trouble, chicken trouble and pony trouble. But in spite of all this, to my gratified amazement, by the end of that summer the garden began to look as if someone were working in it. Two people were, in fact; Mr Guyatt and myself. The only help I ever had with that garden, except for short sharp bursts from friends and family, was from old Mr Guyatt. I don't know exactly how old Mr Guyatt is, but one might say of him that however old he is he doesn't look it. That may sound vague but it describes him very well, and those who happen to know a man of his kind will understand. Mrs Guyatt, I know, is over seventy, and with such black hair, so trim a figure and so youthful a step that we like to say to friends on passing her in the lane, "How old do you suppose that young lady is?" No one has even been within twenty years of the right answer.

The Guyatts kept seventeen cats. We used to discuss these endlessly, Mr Guyatt and I, while toiling in the nettles.

"Terrible on rubber bands, they are," he said. "Can't leave a band about anywhere; we're always short for bottling. Got 'em stowed in a basket at the moment, hanging up in the kitchen rafters."

I thought this a most remarkably original taste for cats, until one day I found Vashti licking her lips with half a black garter of Shaun's still under her paws. After this first experiment Vashti was as keen a rubber hunter as any of Mr Guyatt's seventeen. Bottling bands, small elastic bands, india-rubbers, hot-water bottles, washers; all were quite delicious.

"Now I can understand it with your Siamese," said Mr Guyatt, sharpening his spade (he believed in a very sharp spade, and I am glad to have learned the difference it can make), "they being partly descended from Malayan jungle cats, as you say. Rubber, jungle tree, grows like a weed in Malaya. Stands to reason they could've got a taste for it. But my cats, now. Beats me."

"They got a kitten," he once said to me as we bent above the new raspberry rows.

"What, all your cats? One kitten?" I asked.

"Stands to reason," said Mr Guyatt wheezing gently in the sun, "you can't keep all the kittens seventeen have in a twelvemonth. Thought I'd leave 'em one this time — out of three litters I found this morning, that was. Fat? Clean? You want to see it, with all of 'em washing it and feeding it. Ah, it's a wise child as knows its own mother."

"Or you could just as easily say it's a wise mother that knows its own child, in this case," I said, rubbing my wrists with a dock-leaf. I was always rubbing my wrists with a dock-leaf in that malicious nettle-bed garden.

"It's a fact," said Mr Guyatt. "There's none of 'em

knows, not even me and the missus. But there's no unpleasantness, mind. They all mother it, bless 'em; quite impartial they are. But that does get washed. I only hope," he added, "that I've struck lucky this time and it's a tom."

Making a sympathetic inquiring noise I went on digging.

"Course, that's how we came by the seventeen. Always kept one when they had a litter, see, and I always thought it was a tom. Practically never was. Perhaps my eyes aren't what they were." He blinked them, chicory-blue in the autumn sun. "Don't do any good in the garden, you know, don't cats. I had to give up quite a lot — except me peas, of course — since we got all those cats."

The "peas" were sweet-peas, for which Mr Guyatt was locally quite famous. They were never called sweet-peas, because for Mr Guyatt no other kinds of pea were really worth mentioning. His peas were his pride, and rightly so; usually they bore at least five blooms to a stem and he always reckoned to get a prize with them at the horticultural show. We were occasionally honoured with a bunch of these garden aristocrats at the farm, as part of a scheme of barter that I had with Mr Guyatt in the days when tea was rationed. We never used more than half our ration, being drinkers of weak tea and lovers of coffee as a family. The Guyatts on the other hand liked tea strong and often. I gave him our surplus but, being an independent person, he used to bring me something in return. He knew we loved flowers, my mother and I, and no one knew better than he that our garden had none this first summer.

"I've brought you a few peas," he would say, laying them tenderly down on the table.

Or, "It's clematis this time, off a lovely plant I had from the very best growers. Beautiful flowers, now, aren't they?" holding one up to the window.

"Now don't you two ladies start quarrelling over these roses," he would say, knowing that my mother wanted them for the "granary" and I for the family sitting-room. "There'll be more where they came from, next week."

There was a sad tale once about "me peas." Mr Guyatt also used to put in a few days for Nan's mother, Mrs Wilton, in the garden of Upper Highfield, and there of course he established a row of peas. Double-dug and well trenched with farmyard manure, those peas had everything. With any luck he would pull off a double at the show, with first and second prizes (impartially, since he grew both lots) to Mrs Wilton and himself. But there came a time when he was worried about those peas. They didn't look somehow right, he told us. He was watching them very anxiously because the day of the show was relentlessly approaching and time was short. Then suddenly the whole tale was out.

"It was those twins," he told us wrathfully, meaning Nan's young sisters June and Bridget. "Now you guess what they did, the rascals — they been and sowed kitchen peas down all those rows so they came up simultaneous and crowded all mine out. I went along and told Mrs Wilton straight, I did, and she took my part, what's more. Now that was a shocking trick to play, I ask you! tampering with me peas."

A long time later I remarked to him, "You remember

about those peas of yours that the twins mixed up with kitchen peas?"

"Would I forget it!" The incident still rankled.

"Well, I wondered if you'd mind my putting them into a book? I dare say you may think it funny, yourself, after all this time!"

"You carry on," said Mr Guyatt generously. "At my time of life you get broadminded. But funny? No! That was a *shocking* trick to play. Might've forgiven it, I might, if it'd been anything else but me peas."

This summer we invited, for the first time at Punch Bowl Farm, our very old friends the Webb family: not to stay in the farmhouse, of course, because of lack of room, but in the Old Orchard in their own two tents. Knowing that in further years we would be able to show them gradual improvements, we now displayed with pride the sheer primitiveness of our arrangements.

"Next time you come," we said, "you won't see the well working any more. Perhaps we shall have an indoor what's-it — even a bath. We might have electricity, too, and the beginnings of a proper garden, as well as heaven knows what on the farm."

The Webb boys, Dion and Julian, were fourteen and twelve years old at that time. They absorbed themselves into life on the farm as if they had been reared in the deepest country, though in fact, their father being headmaster of a large town school, their home background was urban. They had a hand in nearly everything that summer. Bill was doing a lot of tree-felling, in the tall neglected hedges, and sawing up the trunks for logs. Dion and Julian, with Shelley, learned to swing an axe and use a cross-cut saw, under Bill's supervision. With Kenneth, their father, they helped to fence the pasture field, Lower Naps, which was far from cattle-proof.

Shelley said they ought to learn to ride. We had a borrowed pony, Twilight, that summer, making three with Tarquin and Red Clover, so Shelley and the two boys would ride off into the Punch Bowl and beyond. Twilight was a handful, only lent to us because she bucked, but when out with the other ponies she behaved fairly well. Sometimes when I was walking on the rim of the Bowl with Eva (the boys' mother) and Kenneth, we would see the three riders, far below us, moving like fairy horsemen across the long green patch that led to the gamekeeper's cottage, or down the bracken-bordered tracks.

During most of the Webbs' camp that year the weather was terrible. The rain scarcely stopped. We issued an open invitation to the farmhouse but the campers stuck to their tents, only coming indoors for an occasional warm-up, a meal or a comfortable evening. One late afternoon, I remember, we looked out after a day of rain and saw the two little tents as trim and spruce as ever. A camp fire smouldered before them, and just above the cooling ashes were suspended Kenneth's rubber boots, feet up, on sticks driven into the ground. Suddenly the same absurd thought struck both Mother and myself. The weather had been too much and Kenneth, despairing, had taken a header into the fire. A more comically pathetic sight than those boots, at that time, I cannot imagine.

My mother and I laughed so much that even the weather seemed less awful.

Cocoa suppers were fun in those days, with nine of us gathered in the farmhouse kitchen. I would make a great panful of foaming all-milk cocoa, and sometimes we would make hot dripping toast or roast apples. Once I even managed a frying of chips. It was all very informal and delightful, and over the mugs and beakers we would discuss the plans for the house and farm and garden.

"When you come next time. . . ."

CHAPTER SIX

Nothing at all to Bill

About this time there was sudden chaos in the kitchen. Bill, home for a while, was tearing up the old cracked concrete of the floor ready for laying it down to quarry tiles (I don't know why these tiles are called quarries, for they certainly are not quarried, being made from cement and sand). The rest of us walked about on a precarious spider-web system of overhead planking, and, the weather being fine, I took the kitchen table out of doors and mixed my cooking in what would one day be the garden. This area, now it was high summer, was so lush in stinging-nettles that we always wore gumboots for negotiating it and outdoor meals were taken to the fields.

Soon, the whole kitchen was lost to sight in the clouds of dust Bill was making in hacking out the ten-foot inglenook fireplace. It was because of the tiles, he said: the floor space went right across into the hidden inglenook, and if he didn't carry the tiles straight through now that he was laying them, they would never level up properly.

People said, "Ah, you'll only regret it! Stands to reason those great old inglenooks were blocked up for a reason.

They'll smoke like anything, you'll see, and you'll all be sucked up inside them with the draught whenever anyone opens a door."

"Oh, will they? Shall we?" said Bill from inside his cloud of dust. "Well, we'll see."

Presently, the clouds cleared, and we could see and breathe in the kitchen. Where usually there is a black oak beam, this fireplace had a delicate wide arch in small rose-red bricks; a delightful surprise that anyone may come upon when restoring ancient houses; and our house was, we now found, much more ancient than we had supposed, a friend having found mention of it in Haslemere Museum under the date 1332. And this was not the building date but a reference to the then tenant, one John without any surname.

In the exposing of the inglenook Bill found, not one, but two brick ovens. There was the main oven, situated in the right-hand side of the ingle and used, we knew, within living memory (this unfortunately had to be demolished in order to make room for a modern cooker later) and there were the remains of a much earlier one jutting out at the back of the inglenook wall into the old wash-house. The roof curves of both these ovens still exist and can easily be seen.

A further discovery, this time in the chimney itself, was an oak cupboard, so old that it fell to pieces in Bill's hands. It measured about eighteen feet by one foot by one foot, and was, alas, quite empty.

The soot-blackened wall at the back of the inglenook was a problem and, in the end, after ineffective attempts to clean the bricks by chipping and wire-brushing, we decided to cement-face it over and distemper the same as the walls.

The kitchen floor required, we worked out, nearly one thousand six-inch tiles.

In retrospect; if we were doing this job again, we would have nine-inch or foot-square tiles instead of the more usual six-inch ones. We have seen these larger tiles in the kitchens of Upper Highfield Farm and Rock Cottage and think they look much more solid and effective for a farm-house than the small ones do. Besides which, they must take far less time to lay.

"Now," said Bill, "I don't think we should put that old kitchen range back — if you can go on managing with the primus for another week or two. What we really ought to have in this inglenook is an Aga. With that, we can have the whole big chimney ceilinged off (you won't see the ceiling) and make an end of all the draughts and dirt. We'll only need a small hole to take the stove-pipe."

This seemed sensible enough; especially as there was a second-hand Aga advertised in the local paper that week. We drove twenty miles in the old Standard, Thunder, to look at it and, finding it perfectly in order, bought it on the spot.

The local Aga agents brought the cooker by road in sections, reassembled it here and installed for us. They also lit it (with charcoal) before they left, explaining that it would take about twelve hours to heat up to full efficiency. There was no reason to suppose that it need ever go out again, but in fact we have let it out once or twice, through human error in forgetting to make it up at night.

Now there was warmth and comfort in the kitchen, where had been frozen feet and streaming eyes and noses. All was clean, where smoke had blackened everything. Cooking was now so easy that it was almost a pleasure: the porridge, made from coarse pin-head oatmeal, was merely brought to the boil in the evening and left in the warm oven till wanted for breakfast; milk puddings were perfection and bread the nearest thing to loaves baked in an old brick oven that anyone could come by.

In those days, with Mother to share the indoor work, I had time for baking bread. We had bags of whole wheat flour sent by post from a Sussex mill, and bread made from that was bread as it used always to be and should be now. It was dark gold-brown, crusty and flavoursome, sustaining and laxative (why not mention the fact, in an age when fortunes are being made from pills and salts to correct the world's most general ailment?). Two slices set you up as four slices of devitalized white bread cannot do. This bread was a world removed from bakers' white, but it was also half a world removed from bakers' brown, that dry rubbery stuff which is never the whole wheat unless clearly guaranteed as such.

I used a useful quick recipe for baking bread in which all kneading is eliminated, and as it was also entirely satisfactory it may not come amiss to write it down.

You need:

3 lb. of whole wheat flour.

2 oz. yeast.

1 tablespoonful of molasses, treacle, honey or brown sugar.

1 dessertspoonful of salt.
Warm water or milk-and-water; about a quart.

Put the flour in a warm basin and add the salt. Mix the yeast to a cream with a little warm water or milk. Add this to the flour, with the sweetening; stir with a wooden spoon, adding warm liquid till the dough is of the consistency of cake mixture; wetter than ordinary dough. Fill bread tins about two-thirds full and leave to rise in a warm place until the dough is almost level with the tops of the tins. Bake in a moderate oven for three-quarters of an hour and take from the tins to cool on a rack. Do not cut for at least twenty-four hours: this bread improves with keeping and remains good for a week.

The kitchen now was always warm and welcoming, no matter how cold or how early the hour. The cats adored the Aga. My Siamese Vashti and Shelley's tabby Patrick would lie luxuriously all over its warm insulated top; I thought nothing of having a cat or two within six inches of my saucepans but Mother never cared for this arrangement very much, being inclined to be allergic to animals.

Bill now turned his attention to the south window. It was ordered and when it arrived, most beautiful with its three leaded sections and metal framework, it puzzled us rather on account of some mysterious holes spaced all around the frame.

"Perhaps you'd better phone the glass-works," said Bill, who was considering the south wall prior to breaking into it for the installation of the window.

"What," I asked the glass-works, "are those little holes for? The ones all round the frame?"

"What holes did you say, 'm?" inquired the glass-works civilly.

Describing them I said they knew those little holes, the ones all round the frame, about as big as a pea?

"Oh, those little holes, 'm; well, really they aren't for anything; no, you just don't take any notice of them, they don't mean anything, you see."

"Not mean anything?" I inquired; then could they perhaps tell me why they made them?

"Well, 'm, I can't rightly say, 'm, but as they don't really mean anything, perhaps it would be best if you took no notice of them, 'm."

Bill snorted. They were a scheme to give employment, he said, but he took no notice of them, and that was the last we ever heard about them. Bill was, at the time, concentrating on the job of propping up the top storey with a skilfully placed upright beam, so that he could cut out the opening for the window below without having the whole house fall down. This one beam, slipped in at an exactly accurate angle under the main ceiling rafter, was gently tapped into a position where it could and did support the entire weight of the south wall and upper gable. During the subsequent demolition, Bill removed two crumbling beams, so rotten that it had been possible to see the light (and feel the draught) through the middle of one. Another upright, which should have been carrying most of the weight of the gable, had rotted loose from its brick base — probably through capillary action of ground damp, as it was quite sound above the bottom inch or two. Cutting out the rot, Bill raised and strengthened the foundation and cemented the beam firmly into place.

Because this particular wall — unlike the eighteen-inch stone walls in the newer wing — was only one brick thin and very damp, Bill demolished it right down to ground level, and then rebuilt it in a herring-bone pattern of two-inch Tudor bricks to window level. The window itself was next set in, with a tiled ledge and sill, and Mother got out the sewing machine for another curtain-making day. From the time it was put in, this window has been such a joy, such a source of never-failing interest with its long view down Barn field, that I cannot understand how families could have lived here for centuries without it. Moreover, the farmhouse is placed so awkwardly that this window and the bedroom window above it are literally the only south-facing windows in the house. Until we came here, the bedroom window was the only one looking in this direction. It was impossible to see into the south garden, or the yard, or Barn field, without going outside the back door or upstairs. Now, we can watch the cows and ponies while we sit at breakfast; and the sun, which never reached the kitchen till the afternoon, can pour in all the day.

The next of Bill's jobs was very hazardous. This time it was the demolition of the old north chimney which, he said, would probably fall down in any bad storm. It was a massive chimney, tall and broad. It would have to come down, Bill said, or we might lose the whole roof and perhaps our lives as well. He would build it up again some day, when he had time.

The problem now was that we had only one short

ladder. We would certainly have to buy a long ladder one day, Bill said, but added that there was no need to rush into needless expense; he would manage somehow. Whisking the kitchen table out he stood it against the base of the chimney. On this he stood a saw-horse and a beehive and on these a section of the incubator, a pair of steps and a box, and on that he balanced the little ladder, binding everything together to one long sapling with many ropes and strings. For the next few minutes until he was safely on the roof we had plenty of time to admire the new kitchen window, trying not to listen for crashing noises in the garden. But it was nothing, nothing at all to Bill; he scaled it as if it were the way he had gone to bed for all his life. It was, he said to Mother afterwards, as easy as falling off a log.

Soon, where the beautiful, dangerous chimney had stood, there was a ridiculous little pot affair sticking straight out from the tiles like a festival cap from a cracker. And Bill had begun on the sitting-room fireplace. At the same time, infected by the fever, Shelley and I attacked the horrible Victorian iron grate in my bedroom. This we found was very hard work. Neither of us had Bill's strength with a pick and mattock — we could hear his battering blows beneath us and the sound of crashing bricks and rubble — so we approached our work with hammer and chisel, levering each brick up from its mortar until the whole grate could be hauled out into the room. This was a beginning, but there was still the remaining built-up area to be tackled, brick by brick.

Exposing old fireplaces is a very exciting, though

laborious and messy affair. It is especially exciting if you have no idea — as with us — of what kind of fireplace you will find. We had all expected a conventional oak-beamed inglenook in the kitchen, and were surprised and delighted when the ten-foot brick arch slowly began to reveal itself beneath the plaster under Bill's careful chiselling. (Later on we were shown, in a local house, an old painting of this kitchen, dating from before the bricking up. In it was shown the lovely pointed arch, a fire on dogs beneath it and the old brick oven in the corner. On a high-backed settle an old woman sat peeling potatoes, and there were chickens round her feet upon the floor.)

In my bedroom I had rather hoped to find another pointed arch. Our neighbours, the Jupes, had found one in a bedroom at Ridgeway Farm, and on the outside wall of our own north gable were the remains of two more belonging to the old wing. The excitement and speculation grew intense as the gap we made grew bigger; but Bill's shout from below came first: "It's an oak-beamed one!"

Dropping our tools we rushed downstairs to look. The sitting-room looked like a refuse dump, but the fireplace was what we were interested in. It was the traditional beamed inglenook, except for an individual and very attractive detail — the widely rounded corners at the back.

Back at our work upstairs, we soon discovered that the bedroom fireplace also had the same wide rounded corners. My hopes for a pointed arch began to diminish; but in the end I was not disappointed to find that this fireplace and the sitting-room one were small and large editions of the same design. The bedroom one is scarcely four feet across, a perfect miniature inglenook with small spanning beam and beautifully built and rounded brick interior.

The excitement over, we faced the inevitable toil of moving mountains of brick and rubble from both floors. We cleared the bedroom by throwing everything out of the window and tidying the garden later. Bill wheeled the mess from the sitting-room out of the door.

"Well, you can't put an Aga in that," people said, wagging fingers at the sitting-room inglenook when the dust had subsided again.

"An idea isn't much use," said Bill, "unless you can adapt it to different conditions"; and he took a large sheet of iron, cutting it and shaping it as one might make a dress. He made with it a simple canopy and ceiling, with trap-door for sweeping the chimney, and he fitted it into the chimney throat with cement. Just as he had forecast, the smoke all went upwards and the heat came out into the room. It was very pleasant to sit by the fire now on early autumn evenings, the sweet-scented logs sizzling and crackling across the dogs and the sparks leaping into the chimney.

One night I remember, before we were used to the fireplace, we removed a large log that was still smouldering at bedtime. Nowadays we leave the fire all night, behind a fireguard, as most country people do, merely adding fresh logs in the morning, but this big log I carried to the garden in the dusk, laying it carefully down upon the middle of the path.

"I think that ought to be safe," I said to Mother. "It doesn't look as if it could possibly start any trouble, though the weather is so dry."

The log sent up its wavering little column of blue smoke and sparkled here and there in the gathering dark. We went to bed. But ten minutes later we were leaning out of windows all round the house, for down at the gate there was such a clanking and clanging as could only possibly mean a fire engine. There it was, polished brass glinting, helmets shining, little bell banging, as the outfit came bouncing swiftly up the drive. We all tore out in dressing-gowns and pyjamas. Bill was away again at this time: could he possibly, I thought, have had a breakdown with the old Standard and somehow got himself home on a full-scale fire engine? I had known him arrive at the gate on a sixty-foot timber tug once, so anything in that line was possible.

"Now what about your log!" said Mother as we hurried down the path. "What a disappointment for them."

At the gate the captain of the brigade called up the path to us, "Is this the right way to Churt?" Behind him the fire engine clanked and roared swiftly backwards and forwards, turning round very smartly in the farmyard and causing all the roosting fowls to affect heart failure.

"Good gracious, no!" I said. "Are you in a hurry? You'll have to go all the way back to the village; there's only a bridle path to Churt from here."

But happily there was no hurry; it was only a practice run, and after a few minutes of cordial conversation at the gate we watched them bouncing down the drive.

"Fancy thinking I would come home on a fire engine,"

said Bill on his return the next morning, having broken down and got a lift in a milk lorry.

Straight away he began on the cellar, mixing sand and cement and laying quarry tiles in all the lovely inset shelves and arches, which were then rough, broken stone. My part of this job was snow-cemming the walls, and later scrubbing out the pink brick floor. I scrubbed it five times before the water really came clean, and then the cellar was a lovely place. There were the hollow worn brick stairs going down from the kitchen, washed of their mud and rosy-pink as only old bricks are; the arched vault where we had found a hundred empty wine bottles, now the place of cream-pans and milk jugs, farm butter and eggs; the deep shelves all round the walls where I stored our bottled fruit; the apple rack Bill had made and the high mullioned window that lit the place so well. We had added another room to the house; and further, that room was now quite dry, as a result of Bill's having put up guttering under the eaves with a drain running out through the garden. Previously, with an unguttered roof, the rain had poured all round the house, running down the outside cellar steps in little waterfalls and flooding first that cellar and then (through a little tunnel in the wall which one can only suppose to have been constructed for the purpose!) the inside cellar too. There was a small pit in the far side of this cellar floor, covered over loosely with a wooden lid; this, presumably, being intended as a final soakaway for the water — after both the cellars had served as short-cut water-courses.

Now, neither tunnel nor pit are required for anything but historical interest. We have had no water in the cellars in all the years since the guttering went up — except, I admit, for a small amount in the outside cellar due to heavy rain running down the steps.

But still there was no main water, no drainage, no electricity nor telephone.

It was through Mrs Fisher's kindness that the telephone came first of these things. She offered us a party line and it was fixed up almost at once. But the water and electricity seemed as far away as ever. Shortages of piping and poles were the worst drawbacks to bringing these services. There was no galvanized iron piping available, we were told, but the water company held out a hope for copper piping at treble the price of iron: and as for poles, there were none available at all. Eventually, feeling that a good water supply was an essential to successful farming, however humble, we agreed to take the copper piping. Bill went out with spade and mattock and shovel to dig a trench for the pipes from the farmhouse to the lane. It was two feet nine inches deep and one hundred and fifty yards long. Had we taken the trench right round the outside of Mrs Fisher's garden the length would have been one hundred and seventy yards, but her permission to bring the water through her orchard saved us twenty yards of piping and trench digging. Bill's fastest time for any one stretch of digging was twenty yards in three hours, in a place where he struck pure sand; but there was another place, over a bank, where he had to dig down to a depth of six feet in order to keep the trench floor straight, for pipes cannot be taken over a curve, and there the work moved very slowly. It took him three weeks of intermittent work to finish the trench, but the saving he made in managing without paid labour more than balanced the extra cost of copper piping.

We argued with the electricity company about poles, offering to fell our own on the farm, but the company didn't care for that idea. My mother then told us a tale of some friends of hers in Suffolk who used their own poles for bringing electricity. Within a year the poles were sprouting, and within two more they were flourishing trees and had to be felled again, which hardly helped anybody very much. But then came an unexpectedly hopeful letter from the electricity company and we gave up our schemes about felling and preparing poles: a new consignment of poles had just arrived, they said, but these were still at the docks. They couldn't say when any poles could be delivered to the farm, with transport as uncertain as it was.

Time passed, and our periodic inquiries brought no further cause for hope: the poles were still held at the docks and transport was impossible to find. This was absurd, said Bill (who by this time had nearly finished digging the water trench), and he wrote suggesting that he should drive to the docks himself, with a full-sized timber-tug, and pick up the poles for the farm. The electricity company, apparently, could neither bring themselves to accept this unusual proposition nor think of anything against it. Accordingly, within three days, we received a letter stating that the poles were now already on the road.

The electricity actually beat the water to the farm. The poles arrived and one after the other they went up, marching over our wild ravine and hilly fields. Workmen in the house were wiring and putting in points. We decided to have no hanging lamps in any of the bedrooms, but only points for table and bedside lamps, because these seemed more in keeping with the rooms.

For the kitchen and sitting-room, where good working lights are so essential, we decided to try fluorescent lighting — then quite a new thing — but we were a little uneasy about the probable effect in so very old a house. The trouble with a central lamp in most kitchens is that one is always working in one's own shadow, except when at a table in the middle of the room. The sink, cooker, dresser, cupboards and smaller tables are almost always placed against the walls. Our cooker is, moreover, set back inside the comparative dimness of the inglenook, and, as the kitchen ceiling is so low, a central light would not have carried far. Except for numerous wall lights, we could think of no arrangement as satisfactory as two five-foot long fluorescent tubes between the rafters. I admit that they don't look very nice. In those days they were not as well designed as now; there were no "swallow-wing" shaded tubes, which might have looked better than the naked ones; and, once having installed the old type, at seven pounds apiece, we have never cared to consider the expense of replacing them.

But I know of no lighting so entirely efficient. Almost shadowless and quite without glare, the light falls into every corner of the room, so that wherever I am working I can see in comfort without having to screw up my eyes or stand sideways. There are now various different kinds of fluorescent lamp, from the cold, pure "daylight" to the soft, faintly pinkish glow of "warm white." We have tried most of them and like the "warm white" best; it is mellow and kind to the eyes, though it does have a queer effect on some colours, making blues appear purple and yellows paler than they are. "Daylight" lamps are harsher and bluer, making a room seem cold, but they render colours as they really are and so are useful in trade.

Apart from their incongruity in old houses, we have one other criticism of fluorescent lighting, and that is the high cost of tube replacements. Ours last for about a year and the price of replacements is, at the time of writing, sixteen shillings and a penny. Three years ago the price was nineteen shillings and sixpence, so it is coming down, but is still much more than the cost of any ordinary bulb.

In use, the fluorescent tube burns as much electricity as an eighty-watt bulb, while giving a distributed light of nearly triple that power.

We had three lamps fitted in the sitting-room; a central fluorescent tube, a standard for the fireside and an angle-poise on my desk. With mixed feelings of sorrow for the loss of the old and joy for the ease of the new, we stowed away the oil-lamps, the flat-irons and the hurricane lanterns, fixing up the wireless again and the other electrical gadgets. Even the cowshed, stable, barn and workshop were now lit by electricity; but still we hauled the water from the well.

CHAPTER SEVEN

Concerning Water

A thing that hampered our initial efforts was our failure to qualify for the fifty per cent Government grant towards bringing the main water. This grant, it seemed, was only available where the farm was already in full production without the main water. But, we said, we can't get the farm into full production without the main water. Previous farmers here had relied on the pond (filled in by us because it was stagnant and a source of mosquitoes) and on a piped supply pumped up by a ram from the stream in Vanhurst Copse. The ram, now in disorder, would have cost too much to repair for temporary use, with main water costs to meet as well. No man, we said, is willingly going to increase his cattle and other stock, and attempt clean milk production, while winding up every bucketful of water, by hand, from a well over a hundred feet deep.

However, it was all of no avail; the committee was unmoved and we did not get the grant. Instead, capital that might have helped to establish a herd for immediate milk production (when the country was crying out for more milk, then a strictly rationed product) went to pay the bill for bringing in the water.

The well-head had been steadily falling into disrepair. The old roller creaked and wobbled as the heavy buckets were hauled up, the brick and timber casing round the bore-hole crumbled to a touch, and the long wire rope was frayed with use and wear. There came a dreadful day when suddenly we lost the whole of the well tackle — heavy wooden roller, wire rope and two buckets — down to the bottom of the well. The old wooden roller-supports had not been able to take the extra weight of the second bucket, which Bill had added to the original single one in order to double the output of water per haul. The crashing plunge of the tackle was frightening enough, but even worse was the following hour or so, with Bill standing straddled above the giddy bore-hole, hauling up the tackle on a grappling-hook. The heavy load, made heavier by the fact that the two full buckets of water were still affixed to it and could not be emptied till nearly at the top, came slowly upwards about a foot at a time. My job was to pull in the slack of the rope from Bill's torn hands, frantically making it fast around a corner-post of the well-house as it came in.

If the well-head could have lasted another week or two this accident need never have happened, and Bill would have been saved hours of salvage work and reconstruction, for, about a fortnight afterwards, we had main water at the farm. After months of counting every pint, water in a tap was to us the greatest wonder in the world. We turned on taps madly and laughed to see the precious treasure run gurgling down our private drain. What did it matter that we still had no bath? No one, any more, ever, would have to wind up water for cows — or for any other purpose whatever.

The drain, constructed by Bill, ran along a pipeway under the north garden to a deep brick tank which he had dug and built in the Old Orchard. After five years of heavy use this tank has given no trouble and required no attention. It has taken all the drainage water from the house, including that from the bath, sink and Bendix washer, and its disposal system is by automatic bacterial action.

This is all very nice and efficient and unobtrusive, I know, but I do feel rather badly about civilized drainage systems as a whole. They are so wasteful. When one thinks of all the humus lost in drains, humus that is badly needed on our farmlands, one cannot help but think how mistaken is the human race in not devising some workable method of returning to the land that which we have taken from it.

I stayed on a farm, years ago, which had an admirable though primitive arrangement, often used in earlier days. At the bottom of the garden (they are always at the bottom of the garden) was the usual little building, only this particular one was a family model, a four-holer with a choice in sizes and heights to suit all comers. Under the wooden seats were placed four buckets, and in the middle of the little shed was a fifth bucket, kept full of fresh dry earth and provided with a handy shovel. The last gesture expected of you, before leaving the shed, was to tip a shovelful of earth into your bucket. That was all. Everything was clean and nice and decent, earth being the best deodorant and cover known to man. The buckets were emptied on to the midden as required and all was duly returned to the land, from whence it came.

The only troubles about this kind of economic sanitation are: its outdoor character, and the work involved in running it. Main or septic tank drainage is so much easier.

In some ways I was a little sorry to see the last of the well. It was a very ancient thing, gone now probably for ever, and it had a certain irresistible air of mystery about it. On cold winter days it steamed like a witch's cauldron, sending up little wisping puffs of vapour. There were always drops of water clinging round it, and ferns beneath the water drops. The brick sides of the bore-hole glimmered with wetness, and down in the far remote depths was a small cold white eye. The water from the well was more delectable than anyone would believe who had not drunk it; cold in summer but no colder in the winter; sparkling, lively, clear.

But the well meant danger and it meant hard work. We didn't want to fill it in, in case some national emergency should one day cause it to be needed again; so Bill made a heavy concrete slab to fit the opening, and on it he wrote while the cement was still wet, "Well over one hundred feet deep."

Did he mean, asked Shelley pondering, "Well! Over one hundred feet deep!" or "Well; over one hundred feet deep," or "Well-over one hundred feet deep?"

The coming of main water banished another pleasant though toilsome feature of the well-water days, and this was the part that water took in the character of the old kitchen. There were the three tall milk churns, so familiar standing in their shining group beside the sink.

They had looked right, in their setting; they belonged to the farm and had probably stood just where they were now for many years, between milking times, waiting to be scrubbed and scoured and turned on end outside to dry in the sun. Each morning the churns had to be filled with buckets from the well, and the clear cold water in them was sparingly used because of the labour needed to acquire it. Water from the churns was used only for drinking and cooking, except in times of drought. For all other purposes there was a bucket frequently replenished from the rainwater tank by the back door. The water in this bucket was hardly the same element you would have thought, as that sparkling lively substance from the well. It was a sombre, brooding, lichen-coloured stuff, goldy-brown and secret, veiling the bottom of the bucket as woodland water veils the bottom of a pool. It was petal-soft to the touch; the lightest lather floated on it, and the various tokens of the seasons hung in it, as if suspended.

In the rainwater you could always find in summer small drowned moths and other ill-fated insects; and in the autumn curled brown leaves from the tall oak on the bank, whose highest branches are higher than the roof-ridge. In winter cat-ice sometimes tinkled against the pail-sides, or the pale tea-brown water might be clouded with newly fallen snow; storm-loosened moss and lichen slowly revolved beneath the surface as the bucket was swung and lifted to the sink. But in the spring pleasant signs of returning warmth and colour came into the kitchen with the rainwater, and these were small and airy, floating lightly on the surface: damson petals, whiter than milk; tawny-golden sheaths of beech-buds; rosy petals of the peach against the wall, and, later, the larger pink petals, like seashells, of the clematis Montana Rubens that covers the well-house roof.

We could have continued using the rainwater after the mains had been connected — it was pleasanter and far more efficient for all washing — but somehow, in a busy house, such things are allowed to lapse: especially when hot water is in the tap, for the soft rainwater must always be heated as required. Bill sometimes talks of plans to catch and conserve all our rainwater, so that it could be diverted to the boiler whenever supplies permitted. We knew a house in Sussex where this had been done. But when I think of the vast tank (about three thousand gallon size) which must first be constructed, underground, I know that we shall continue as we are.

Now that we had the main water, Bill suddenly said that it was time we had an indoor w.c. Of course we encouraged Bill at once, because he had a great many ideas about the house and land, and one was soon replaced by another unless he made a positive start in good time. We managed to get him into Guildford to help in choosing the best kind (we had a green one with black accessories), then when we saw him safely started on the plumbing we began to think of all the things we should miss about the little place down the garden. For one thing, quite a lot of life's humour would be lost to us at once. The way the fowls all followed one for instance; it was no use pretending to go for a quiet stroll or a glance at the weather. Besides, the poultry clustered round the doorway waiting cosily for one's return — a useful clue for folk indoors, because it often saved a wasted journey.

From time to time we debated other means of displaying the "engaged" sign to the household, especially after I began seriously to tackle the garden and so shut all the poultry out. It was essential to have some kind of signal, I suggested, because the path was always long and sometimes wet, and futile jaunts were silly. My mother suggested hanging out a flag, but the rest of us thought this rather obvious to any passer-by. We never had any passers-by, said Mother, but we didn't care about the flag. For myself, I just left the door open, which was an admirable idea because The Place was built discreetly sideways to the house and, moreover, there was a nice view through the doorway.

"If it's a view you want," said Bill, and took his keyhole saw and cut an openwork design in the door at what he judged to be eye-level. It wasn't my eye-level, but I preferred the open door. Anyone from the house could see that it was open, but that was all they could see; and an open door, I maintained, was more discreet than a fluttering flag.

Another feature of The Place had been its squirrel, who used to store his nuts — our walnuts — in a hollow in the wall beside the seat. You could usually count on finding a few good ones in season. The cats had no inhibitions about privacy, and even with folks who shut the door they were not deterred from popping in and out on social visits, the already wide space below the door being trebled by the deeply hollowed step, worn dish-shaped by the feet of centuries.

But these nostalgic thoughts were purely frivolous. As soon as Bill's masterpiece was finished we accepted it with delight. At first we were not above the childish sport of dropping in bits of twisted paper, pushing the flush, and then rushing downstairs and out into the Old Orchard to peer down into the deep brick drainage tank for the arrival of our twist. It was wonderful to think that the thing really worked; even more wonderful literally to see it deliver the item entrusted.

"I always wanted to work one of those cash-delivery systems in shops, when I was small," I said, "but this is much better fun."

Bill, however, sealed over his tank and that was the end of that.

CHAPTER EIGHT

Gun-cat, Plough & Garden

The long delightful summer slipped to its end, and with it almost all our leisure hours. It was Michaelmas, the beginning of the farming year, and the fields we had lent were now cleared of their harvest. They lay there rough and stubbly, playground of hordes of rabbits from the miles of National Trust land marching level with our own. This winter the fields were in our hands, hands that had never held a plough. We put ourselves wholly under the guidance of the Agricultural Executive Committee and, while awaiting their bidding, we went out and shot the rabbits. Then, as there was a great company of rabbits, we invited others to join in.

A gun was a new thing in my life; I don't think I had ever held one before. Now I had the use of a little folding 4.10, a poacher's gun, and except for one evening's target practice with Bill I was left to get on with it as best I could. Certainly I had never killed anything before, but rabbits were different. They were a threat to food production; soon we were to see how big a threat they were.

"Chap who lived here before the one before you," we were told, "sometimes made more money out of rabbits than what he did from farming."

My only comment now would be that he must have been a good shot. More probably, says Bill, he worked ferrets and laid down traps and wires. We can't use traps and wires ourselves because of our hunting cats, though I have put wires out in the depths of winter when the cats don't go very far from home. The wires were never satisfactory; they caught rabbits all right, but the foxes almost always got the rabbits before I did. I tried going the rounds of the snares at all sorts of times, from night inspections by torchlight to very early morning ones, but hardly ever found more than a rabbit's head or a bit of fur in any of my wires. One morning, walking round on freshly fallen snow, I could see plainly the tracks of foxes that had done the rounds before me; the tracks led straight from one snare to the next, taking in every one I had set, and again there were no rabbits for my bag. I gave up setting snares.

I quickly found that shooting with a 4.10 was more a matter of stalking than of marksmanship. The range is barely twenty-five to thirty yards, and that is very close to a rabbit. Probably my natural silence in getting round outweighed my failings as a shot; anyway, I usually got a rabbit when I went for one, even when it meant creeping on my stomach up the whole length of a stubbly field in heavy dew. I was no sportswoman, always aiming for a "sitter" but out of purely humanitarian reasons; I wanted to be sure of killing the rabbit, not sending it wounded to its burrow. Neither of us shot for sport in any case, but wholly to rid the land of some of its swarming populace of rabbits.

The first time I ever shot at a running rabbit I was shaken into doing it. I had been creeping up as usual to an unsuspecting sitter in Lower Naps. Expertly, I was keeping very close to the hedge, moving silently and slowly. Then suddenly I was shocked by a loud double report, just the other side of the hedge and only a few yards from me. My rabbit cocked its ears for one paralysed moment and bolted for the hedge. I cocked my gun after a similar paralysed moment and shot the rabbit. The amazing thing is that I shot it dead. Picking it up by the hind legs I swaggered round the corner of the hedge.

"Oh, it was you, was it?" I said, coming face to face with Ron Wassal the village baker, one of those who helped us against the rabbits. Well he'd be jiggered, said Ron, he had no idea there was anyone else about, I didn't half give him a jump.

"You gave me one, too," I told him, dangling my rabbit. "Didn't you get yours?" I asked innocently. "Too bad, better luck next time."

Ron was the kind of obliging baker you don't get in a town. He would come up to the farm at all hours, early and late, and he would always do a job, such as splitting and carrying logs for our big inglenook, in return for the pleasure of the shooting. He would ice a birthday cake for us in the kitchen with beautiful skill, or read a bedtime story for Shaun, or make special horseshoe-shaped loaves in the bakehouse for Ann and Shelley while they watched.

Vashti (who had now collected the diminutives of Cat or Catto, derived from respectfully addressing her as O, Cat) became at first my biggest handicap to shooting. She would come too. It was almost impossible to fox her. Sooner or later, when I was thinking she was safely shut indoors, I would hear that unmistakable Siamese howl close behind me and, turning, would see her dark triangular face forging swiftly through the grasses in my wake.

"But we did shut her in, Mother," Shaun or Shelley would say. "She must have got out of a bedroom window or something."

"Very well, Cat," I said, "if you must come, you must earn your company"; and I set myself to teach her to retrieve, for most of my rabbits made one last bound into the depths of some impenetrable gorse or bramble thicket, and I used to get much scratched with fishing them out myself.

Catto took to the gun as a war-horse is said to take to battle: *Who saith among the trumpets, Ha, ha; and smelleth the battle afar off.*

She adored gunfire. Diabolical blue squint flashing, black whip-tail flicking, she would snuffle the air like a tiger. (Was it for gunpowder? the old battleaxe! or for blood?) In one day she learned the rudiments of retrieving better than we had ever instilled them into Glen, and she would snake her way into the overgrown double hedges, snorting and sniffing, hackles up and tail fluffed out like a squirrel's, then slowly drag out to me a rabbit sometimes bigger than herself.

Her only disadvantage as a gun-cat was the breaking-out under excitement of her early talkative nature. She would loudly discuss the prospects, the weather, the route. Not always, I haste to add, or we would never have got anything, but often enough to make shooting with her rather a lottery. But I was very proud of her; she was probably the only trained gun-cat in all England, and though she was not very much trained she certainly was a gun-cat. She was also perfectly well able to catch and kill rabbits in her own right, probably accounting for many more than I did with my gun.

However, I once nearly caught a fine cock pheasant with my bare hands. Luckily I had Shaun for witness or my family would never have believed me. Shaun and I had gone up into Lower Naps one winter's day (this farm, and indeed all this district, is full of such apparent inconsistencies; one can "go up" to both Lower Naps and Lower Six Acres, as well as into Little Bottoms and Highcomb Bottom; it is all a matter of relative heights) to look at my rabbit-wires which were mostly set around that field. Seeing something moving in the hedge I went towards it, suddenly realizing with astonishment that it was a cock pheasant. Behind it and buried in the hedge was a short stretch of wire-netting left over from long-ago days, and back and forth against this netting the bird shuttled.

"You stay at that end," I said to Shaun, "and I'll try to reach him."

I stretched my arm into the hedge, was severely

scratched by brambles and could only just reach the pheasant's tail. This I firmly grabbed, but the bird was spurred to further efforts by my touch and with a loud whirr he was gone — leaving in my hand his splendid tail!

At home, I stuck the feathers up beside the inglenook, and was much teased by my family.

"But she did, she really did!" said Shaun.

The next day, when out with Catto and the little gun, I saw the fine cock pheasant once again, there was no mistaking him, flying erratically through the pines without his tail. I had a pot-shot at him — it would have been a proud moment when I added him to his tail — but I was glad I missed. He deserved his freedom, having escaped me twice. Doubtless by now he has a new tail and is very careful about hedges with old bits of netting buried in them.

A thing my gun-cat cannot stand is high-pitched music. It hurts her ears, I think. One evening by the fire Shaun began to practise notes on his first mouth organ. Catto and Patrick were sitting one each side of him at the time. The first note made Vashti's ears start twitching, and the second laid them flat. Rising, she advanced across the rug towards poor Patrick, who was dreamily blinking at the fire, and dealt him a sudden swift blow.

"Take that, you ill-bred animal!" she obviously said, her blue eyes squinting angrily at him, "and never make such a noise in my presence again."

With which she swept out.

* * *

Bill, who had given up his job, was now faced with tackling the farm. I hadn't said much, because I thought I was fully occupied as things were, but it is surprising how much more than a bushel will go into a bushel pot if you really cram it. Just then I was making renewed and valiant efforts in the garden. I would have a garden, a beautiful garden, a proper garden, I said, at all costs. I was sick of this great rubbish dump; of having nowhere to sit and write; nowhere to take meals; nowhere for a family outdoor room at all, except the fields and heathland. The brave beginnings that I had made with Mr Guyatt now shamed the rest of the garden and I was committed heart and soul to bringing order into the desolation. My mistake was in doing it all so openly and in saying so much about it. Bill, single-handed in the middle of thirty acres, quite rightly offered his opinion.

"If you've really got all that time to spare from the house and the children and the writing and the cows, I think it would be much more practical if you spent it learning to plough, and in hedging, and scything ragwort, and things like that."

Very well, I said recklessly, so I would. I would do all that and still have my garden. I would, I said, have a labour-saving garden, mostly lawns which a mower would keep in order and not very many flowerbeds to take up hours of care.

Now there was no looking back; I had indeed put my hand to the plough in both senses. Learning the mysteries of the second-hand tractor I ploughed till my hair grated together with dust and my nostrils stung with exhaust fumes. Sometimes Bill drove the tractor and sometimes

I did, but one of us always had to walk behind the shilling horse-plough which we had bought at the sale, struggling to keep it from lurching in the stony furrows, from building up its own shifting mountain of bracken roots and from finally capsizing with its helpless shares stuck into the dusty air. Our furrows were of all sorts, sizes, shapes and in all directions; but they grew better. There came a time, impossible though this seemed at the beginning, when ploughing was like sailing one's own boat; the forging onwards on a clean straight line, the sun and wind so strong, the brown earth lifting, turning and breaking like a long wave under the bows of the tossing little horse-plough. Ploughing with a tractor, I found, one is mostly looking backward. Once the first furrow is cut, its sharp wall holds the tractor wheel fairly firmly to its course, but behind you anything can happen — specially with a shilling horse-plough. In rough ground — and we did plough some rough ground — we constantly had to call a halt for spading out the bracken roots that built up underneath the plough. Now that we have a two-furrow tractor plough and the fields are cleaner, we remember those days, as pioneers harking backward.

Ploughing was not all. I could not spend time in the garden with a clear conscience now unless I spent at least twice as much time on the farm. I helped burn bracken and gorse with flaming torches, scythed and pulled ragwort, layered hedges, harrowed, cultivated, rode the seed-drill, scattered fertilizer and reset roof-tiles, and also made the garden.

These things were only possible, of course, because of the practical support of my mother. Without her cheerful and efficient housekeeping then, we should still have been besieged by nettles now. But if only, oh, if *only* I had heard in those days of sodium chlorate — the wonderful weed-killer which will destroy for you every living plant that is in the ground, leaving a garden clean and fit for planting within six to nine months. Now I cannot use it, because of treasures in the garden, but then there was nothing at all but couch and bracken, nettles and ground elder. I dug and hauled out what I could, and the rest I dug in. Somehow I achieved a kind of level (as with the fields, it was relative), only calling on Bill for help where some heavy job of stone heaving or old gooseberry bush removing was too much for me.

Slowly the garden grew, right on through the winter. I planted roses, irises, clematis, Magnolia Soulangeana on the north lawn, honeysuckles (Halliana and Americana, for successive flowering, against the north and east house-walls; both have been spectacularly successful), and herbaceous plants and alpines without number. The large central areas, roughly levelled as they were, I sowed to lawn-grass seed, raking and rolling and hoping for the best. In fact, the worst really happened, as it must with hastily-made lawns; my lawns are now as full of clover, dandelion, plantain, creeping buttercup and coarse grasses as they are of the grasses I sowed. They look all right at a quick glance but they are very hard to mow, the mower knives slicing through the coarse stems like a bacon-slicer through old ham. The nettles and ground elder came up cheerfully with the grass-seed, but I was not worried about those; three or four mowings were the death of them and they have never showed face again on any of the lawns. It is the creeping weeds that so ably survive the mower; and, of course, the flat rosette type such as plantains. I don't mind the clover at all, in fact I encourage it; it is low-growing, dense, and very green; it is springy to walk on, easy to mow, charming in flower and beloved of the bees; but you have to be careful with it near the flowerbeds, for it has wandering habits.

Because of not knowing about sodium chlorate, perennial weeds have been my worst of all troubles in the garden; especially that diabolical trinity, couch, ground elder and nettles. So firmly entrenched are they in all the old paved and cobbled paths that there is no eliminating them; they creep continually into beds and borders and even grow in cracks of walls. A cautious use of mild weed-killer holds them back a little, but I cannot be properly lavish because of garden plants whose roots might be affected. But though they were my worst trouble they were far from being the only one. I was to suffer much tribulation in the garden from invasions of cows, ponies and hens; indeed, I still do. One gate left open for one minute and five years' labour can be ruined. I have been in tears over livestock in the garden, but one becomes almost inured even to that disaster in time.

I spent delightful hours with nurserymen's catalogues, writing lists, crossing off lists and writing more lists. ("Think," said Bill, "of all the lime that would have bought for the fields, the muriate of potash or the tractor-oil.") My catalogues became a family stock-joke:

"Mother's homework," "Ma's bedtime reading," "Give you three guesses what she's reading now." "If it isn't those it's a gardening book, or that bit in the *Sunday Times*."

I began my gardening career, as most people do, with a variety of seeds and bulbs, numerous perennial plants and far fewer trees and shrubs. One gradually works upwards, I find. It has been said that everyone who perseveres at gardening for long enough in the same place becomes a tree-and-shrub gardener in the end. It is the top of the gardening ladder; but in those days I was influenced by cheapness, quick effect and (as I thought) easy cultivation. In fact, of course, the trees have both the first and third of these advantages if you take a long view. They are far cheaper *in the end* because, though costly at the start, they increase in size and beauty yearly, never needing the annual replacements of so much herbaceous stuff; and they are a great deal easier to grow, providing one has first carefully considered their situation (with respect to soil and wind shelter, shade or sun, etc., as individually required) and seen that the original installation was well done.

The quick effect can still be achieved by interplanting with lesser things, to be removed later on when space is needed by the growing trees and shrubs, but I don't do this myself; partly for lack of time and partly because I prefer now to spend all available garden money on more trees and shrubs. Perhaps I am getting too middle-aged to bother so much about quick effect, anyway. The time is past when everything had to happen immediately, and the time has come when there is fun in long anticipation.

As one gets older still, I suppose, the desire for quick effect returns, so that, of the brief years left, none shall be wasted.

And all this time I was writing, of course, for until the farm began to pay my books were to be our main source of income. Our children and their friends had always been the chief characters in my stories; a little disguised, perhaps, and with changed names, but easily recognizable. I wrote about real places, people and things; probably because in this way "liveliness" is best achieved, or just as probably because I haven't the imagination to create everything from nothing. Before we came to Punch Bowl Farm I had written against the Romney Marsh background of my childhood home, but now I was able to start a new series of books altogether. The characters were still our own children and friends and animals as before, but now they appeared as a different family living here at Punch Bowl Farm. The farm and surrounding countryside went down exactly as it was, every field and track and building, every animal and creature on the farm, right down to the beautiful mad hen, Phasian, sole survivor of a fox-raid, whom Shelley lovingly adopted, and the cowshed mice and Shaun's snails.

All writers, if they have been writing for long enough, receive letters from total strangers. Mine, as one would expect, are mostly from children, but they are not always so. One of the nicest I ever had from an adult was from a lady who wrote to say she thought I would like to know that her maid (I hardly knew that anyone still had one) always read all my books as they came into the library, and had said, "Ooh, I think she's lovely; I like her better than Denise Robins!"

As often as not, the children who write to me ask, "Is there really a place called Punch Bowl Farm? and are Peter and Lindsey and the others real people?"

One of my greatest pleasures is in being able to say: "Yes, it is all real; Dawn and Duchess, Tarquin and Red Clover, the children, the farm, the wild valley, the deer and the Devil's Punch Bowl; and the map on the end-papers of the books is the scale map of our farm."

CHAPTER NINE

Time Out of Mind

I think there must always have been people living here, where Punch Bowl Farm now stands. How is one to put a date to a farm where the existing house is of two different and distinct periods, where foundations of other older houses or parts of houses are constantly being excavated, and both early stone age and Roman implements have been ploughed or dug up?

Archaeologists who have looked at the place have admitted that they are baffled. The newest part of the house, the sitting-room end, is probably seventeenth century. It is built of Bargate stone, very solid and strong, with tall windows and only a little timbering in the high spacious rooms.

The kitchen end is probably sixteenth century or earlier. The style of building here is entirely different, being based on an oak timber frame filled in with wattle or brick. The timbers are, as was usual, largely wood from old warships, the dowel-holes, slots and grooves in unexpected places being evidence of this. The ceilings in this part of the house are low, the windows long and low, and the bricks are all the lovely small rosy-red ones of the period.

From the garden one can see the rather clumsy alliance of these two sections of the house, a clear perpendicular line in the walls showing where the new wing was built on to the old. Except for the strange swooping line of the roof-ridge, there is little to see of the joint above the eaves, for since the building of the "new" wing the tiles must several times have been removed and taken smoothly over the whole roof.

These tiles now have a more ancient look than almost any other part of the house. They are, one might say, more like a natural outcrop than a man-made roof. Each separate tile is different from the rest, with just that small difference in size, colour, warp and placing that makes the whole so lovely. Each is encrusted with its own particular arrangement of mosses and lichens, each catches and throws the sunlight in its own way. But looking at the roof as a whole its unity is expressed in flowing lines and dips and waves, just as the unity of a flock of starlings (each bird as individual as each tile) is shown in concerted patterns of flight.

Under the chimneys the moss grows greener and thicker in long bays swooping down the roof, for here the tiles are sheltered from the downrush of the rain, carrying with it, as it does in heavy storms, cushions of moss to throw upon the paths below the eaves.

Looking downwards, one sees the brick arch in the north gable, the older part of this house, where was the inglenook of a room long fallen down. Here, where the path runs down to the gate, were once the fireside chairs, the cats and dogs, the ordinary comfortable things of a family hearth. Where the yellow of my winter jasmine lights up this north wall, the brighter yellow of hot flames lit a room that no one now remembers.

We were fortunate when we came here in getting to know an old lady of the village, Mrs Warner, who was born in this house more than sixty-five years ago. She sometimes talks to us about the house as she remembers it when she was a girl here. No one, she says, knows anything at all about the old north wing. She never heard her parents or anyone else mention it, and it was, then, the same forsaken-looking jumble of footings that it is now. The manner of its fall had been a mystery through all her life, she said.

Now we are practically certain that the old wing met its end by fire. We think it was pulled down, while burning, in a successful attempt to save the rest of the house, and we think it may have happened about three hundred years ago or earlier, if the "new" stone wing was built just afterwards, as seems fairly likely. Evidence of this fire is plentiful for those who dig and poke and peer. All round the old wing site we have unearthed broken bricks and stones with marks of great heat on them.

In the house itself there are also signs of burning, and these are all in the rooms that were joined to the fallen wing. Charred faces of beams, blackened corners of door-posts, burnt areas of floor against the wall; all are there, easily to be seen.

More evidence was unearthed when Bill was digging the trench for the water pipes. Beside the track, where the pipes were to go, a long humpy line was found to contain nothing but old building rubble beneath the covering grass, and much of the rubble bore marks of deep burning. Here, we imagine, the rubble was carted after the fire and thrown out beside the farm track. Perhaps the farmer planned to build a wall with it down the track one day, to stop his horses and cattle from straying out whenever the gate was left open. But the farmer, being a farmer, never quite got round to doing it; there was always something else more urgent; a cow had calved, the hay was just right for cutting, the heifers had got out or the tilth was perfect for the seed.

Or did he, perhaps, intend to make his track more firm and solid with the rubble? Did he hope to build a spare barn? down there where our tall Swing Tree was probably a sapling left by him to grow because it was a straight and shapely one.

People from the south (existing) side of the house went through to the old wing by means of a door beside the inglenook. That doorway, for centuries bricked up, can still be seen quite easily from the garden. From inside the house, the plaster hides its presence, but we know exactly where it was — at the end of the larder leading off from the kitchen. Our larder was once a passageway to the old north wing. Sometimes, when I go in there, I think about all the people who have passed that way but not just to turn round and come straight out again as I do; they walked straight on, through the door, and out — to what?

Sometimes I try to imagine what I should feel like if, on going into the larder one day, I should see, not the shelves and bins and benches but a clear swept passage and at the end a door. Opening it, would I step right out of time and see the old wing as it was? With what sort of rooms, windows, furnishings, people?

This sort of thing only happens to ordinary people in fiction (I made something like it happen in my children's book, *Spirit of Punch Bowl Farm*, and enjoyed the writing enormously), but it is the only way by which anyone now could know the mysteries of the old north wing. No records exist, so far as we know; even the deeds of the farm only go back to 1926, for some incredible reason, and one of the special conditions of the sale was that the purchaser should not "raise any requisition or inquiry as to the earlier title."

A strange thing Mrs Warner once told us was about what the children always call The Treasure. All ancient houses should have a secret room, a fallen wing, and a treasure. Ours has all three.

When she was a very small girl, Mrs Warner told us, her mother received a letter. Any letter in those days, on such a lonely little farm, was an event, but this letter was from France and from an unknown correspondent. He had, the writer said, a plan of this farm showing where something valuable was hidden. Might he come to England, to the farm, and see if he could recover it?

"What did your mother do?" Shaun asked, his eyes as wide as the eyes of youth always will be at the thought of hidden treasure. "Did she write? Did he come?"

No, said Mrs Warner, no, her mother didn't write. She just took the letter and threw it into the fire saying, "No good can ever come of that!"

"What?" cried Shaun, "and lost the Frenchman's address and everything?"

Lost them completely, said Mrs Warner sadly.

"And did he ever write again, the Frenchman?"

No, he never did, so far as she knew. And as to what the hidden thing was, or where it was hidden — how could anybody say, now?

We have never found anything here that might be called valuable, but we have found much that was interesting. Before we came, some Roman pottery was ploughed up in Yew Tree field and is now in Haslemere Museum. We ourselves have found small shards that we think are Roman, but the oldest things we have unearthed are a stone age scraping knife, beautifully chipped and shaped from flint; a hammer-stone, and a boiling-stone which was used for boiling water in earthenware pots, by first heating the stone in the fire and then dropping it into the water-pot — these stones are smooth, round, glazed and cracked from long use.

Among less ancient discoveries we have found an iron pot-hook; a brass box of clock-wheels that worked a fireside spit; a beautiful little hand-made gold-brown glass bottle, flask-shaped and with a loop for wearing it on a belt; an ox-shoe, and a wrought-iron rein-guide made, we think, to bolt upon the beam of a plough, and with one rein-ring only left to it.

The foundations, of course, are for ever cropping up. Wherever one thrusts a spade into the ground it seems to me that foundations are likely to be met with. I should think that building has been going on here for a thousand years or more. As fast as one part of the old house aged and fell in ruins, another part was thrown out somewhere else.

To my certain knowledge we have struck foundations in at least four distinct and widely separated places all round the house. There is the main ruin, that of the burnt out wing, at the foot of the north gable; there are footings struck by Bill when putting in water pipes, outside the garden gate; there are others found when an arch was being made to span the path for roses, clematis and wistaria; and there are yet others which I found in the yard, just outside the back garden gate, when excavating for stone to make a path. None of these, except the north wing ones, seem to have any direct connection with the existing farmhouse. All are very solid, heavy and deep, and are bound with old lime-mortar.

We have hunted through old volumes of Surrey history in the vain hope of discovering some clue about the earlier days of the farmhouse; some drawing, perhaps, of how the old house used to look. From the foundations left, we cannot even build up the vaguest picture of the structure as it was. No one here in the village could possibly remember, for it was "time out of mind," and no one, so far as I know, has ever heard. Cobbett rode this way, as he mentions in his *Rural Rides.* Coming down our hollow Rocky Lane and into the village he must several times have passed our gate and seen the house, though he never mentions this farm. But this, of course, is no real loss for his time was too close to our own for the house to have looked very different. A century or so will have passed over it leaving hardly a mark, except that once (thirty years ago or less) it was entirely covered with ivy, we are told, so that it looked like a great green hummock with chimneys sticking from the top of it.

On five ancient farms around this quarter, the two Ridgeways, the two Highfields and our own, the passing centuries were beginning at last to take their toll. Houses over which four centuries can pass almost unnoticed, may crumble at the passage of two more. These five farms were crumbling. For all of them this century was a time for dissolution or restoration; they had come to the final end of their natural tether. All five now have new leases of life; all, for the first time I believe, are now owner-occupied and thus personally loved and cared for. In the last ten years, only, they have emerged from generations of decay and set out upon new generations of usefulness, lovingly repaired, carefully restored and strengthened.

CHAPTER TEN

Hard Winter & Some Siamese Cats

That second winter began with a flourish; Bill and I ploughing our own land with our own hands and implements, sowing our first crop of corn, and planting the turnip acre (now known as Little Orchard) with the sixty selected orchard trees for the house. Bill and the children and I ploughing and harrowing Yew Tree field (Shelley taking her turn with the tractor and harrows though she would have preferred a team of horses), tearing from it mountains of couch roots which we fired with a blowlamp, Bill once laying a long snail-trail of roots across the field for Shaun to fire, making a smouldering snake of it. Bill and his brother Sam and I planting the whole of this field with two hundred and seventy-five Cox's, Laxton's Superb and Sunset apple trees for a future commercial orchard.

It was not on the whole a winter noted for hard weather, and when the frost did come, the planting was all done. Then came the snow, a fortnight of it, deep and white as snow is never seen in towns. The sunk lane to the village was cut off to all wheeled traffic and we fetched supplies by sledge. Shelley, unable to go to school, employed herself with making leather sledge-harness for Glen from an old cart-horse head collar found in a corner of the stable. She tried it on him as she worked, as a tailor fits a jacket, and then she harnessed him to the sledge and taught him to pull it on the snow.

The dog had always known she was his mistress and he was pleased to do anything she wished, except not trail her when she rode to school; many times she had to turn and bring him back, having suddenly found him at her pony's heels when far out from home, though we had kept him in the house to give her a fifteen-minute start. He was very pleased to pull the sledge for her, and did it with a proud enthusiasm, though Shelley always helped him with the hauling coming home, uphill and with a loaded sledge. Later on I learned that it is unlawful in this country to use a dog for hauling. I wish whoever made that law could have seen how Glen enjoyed himself as a sledge-dog. Surely all sensible dogs are happier when they have some useful work to do? I think that sheepdogs, guide-dogs, police-dogs, gun-dogs, live a fuller, more contented life than do mere pet-dogs. They are certainly nicer to know.

In the time of deep snow all our animals were kept in sheds and stable. The ponies, used to wintering out both night and day, particularly hated this; and even the cows, who were usually out only in the daytime, grew bored with the long captivity. For us there was much extra work in water-carrying, straw-carrying and mucking-out. Deep paths were cut in the snow to all the buildings, but between them it lay knee-deep, and deeper still in sheltered places. It hung on wire-netting, so that when the wire was knocked the snow sprang out in showers. It lay on the little orchard trees in tufts as if the trees were in miraculous winter blossom. On taller trees in the Old Orchard it lay so thickly and yet so lightly that Shelley and I, trudging under the branches, could make small personal snowstorms by just slapping on the trunks. Gates pushed open left a smooth arc cut in the snow where the bottom bar had swept it. The telephone lines were down, and the farm was cut off from the world except for the wireless and the journeys of the sledge.

Bill and Shelley cut a circular path round the farmyard for exercising the animals, and round this Tarquin and Red Clover and the cows were walked whenever time could be spared to lead them; but I think this only served to whet their appetites for more exercise, as soup tickles the palate of a hungry man.

When at last the snow had thawed to ankle depth we let the four of them out in their respective fields for an hour's freedom. No one at the farm did any work for the first half hour or so, the animals were much too spectacular to miss. First the ponies, pawing the ground, throwing up their heads to draw in great deep breaths and blowing outwards with loud snorts like stallions; then, shaking their manes, madly galloping straight up the steep slope of the Old Orchard to vanish through the gate to Hanger field. Down Hanger they pelted, manes streaming, tails held straight up and flowing out like banners; into the Old Orchard again they would race at the far bottom corner, small Clover stout and already blowing, Tarquin very black and magnificent against the whiteness of the snow.

They slid to a standstill, snorting again by the pond. To taste pond-water, rich and flavoursome, after bucket-water in the stable! The ice on the pond was still firm and Tarquin struck it with his hoof. Broken ice and water flew into the air and over his hoofs; he lowered his muzzle to the jagged hole and drank deeply, then stood aside for Clover.

Away they went again, pounding the snow into a grey slushy path, up the Old Orchard, through the gate, down the lovely rounded hill of Hanger field and into the orchard again at the bottom. Round and round they went, like merry-go-round horses, till they were tired of it, and then they stood by the gate at the end of the track and waited for the baker's van. There might be a loaf that had fallen in the mud somewhere; it had once happened and the ponies never forgot.

The cows were let loose first inside the yard, as they were both in calf and we did not want them to tear about Barn field. They wore their energy off by tilting at the dung heap. If you had been there, and were able to imagine yourself in place of the dung heap, you would know what it felt like to be gored and trampled by a really excited cow.

Duchess was the funnier of the two, because Dawn in any case had no dignity, but seeing Duchess charging at the dung heap and gallumphing about with drapes of straw on her horns was the same as seeing a human dowager sliding downstairs on a tea-tray. Even a cow *en fête* can tire, and the glorious climax came when

Dawn rushed right to the top of the dung heap, followed by statelier Duchess who, becoming bogged part-way, failed to scale the final peak. The sight of her backwards slither, finishing up on her behind in the yard, was one of the highlights of my life. There she sat, in the classic attitude of cats by the fire, her great udder spread fanwise in the snow and her small curling horns bedecked with cast-out straw like an Ascot hat with plumes.

When the snow had gone we began to look ahead towards the spring; but what was left of that winter was to make it the worst that we had ever lived through in the whole of our combined lives. The troubles began with Dawn, the heifer, slipping her first calf dead before its time, the vet saying that the cause was probably contagious abortion. We had thought we were so safe when we bought Dawn, because we bought her at a herd disposal sale.

"It isn't like an odd cow in the market," I had said to Bill. "You always wonder why they're selling; but this is the whole herd going. We're sure to be all right."

Now, of course, we began to wonder why the whole herd had been sold; especially when we remembered that all the animals were catalogued as having been inoculated with S19 against contagious abortion. Few farmers, in our experience, bother about this inoculation unless abortion has already hit the herd. Some, indeed, say that the inoculation predisposes to sterility, but there is, I believe, no proof of this. It was because of the inoculation that the vet was unable to say definitely whether the trouble was due to *Brucella abortus*, for once the animal has been inoculated all blood tests show a positive result. A further complication with Dawn was that, some days before calving, she had turned sideways in her double stall and had slipped and fallen upside down into the low manger. Shelley found her there, legs stuck straight upwards and barrel-body firmly wedged. We got her out and on her feet again with ropes tied round her legs and horns, but she was very shaky for a while, and so were we.

We began to look at Duchess, now in calf again, with some anxiety; and also at the ponies, for the *Brucella abortus* can attack horses with fatal effect. In their case the germ produces an outbreak of practically incurable abcesses known as fistulous wither, but our vet assured us that this was very rare. Many farms had abortion but did not find the infection pass on to their horses. He had, he said, seen only one case in all his years as a vet. But Tarquin did not escape it.

For a month Shelley and I nursed him with faithful labour in the stable, cleansing the pus from his shoulders, clipping all round the places and dressing him with special ointments. Vaccinations were tried and all that could be done was done, but after four weeks we were told there could be no recovery and Tarquin was put down.

From that day I made over to Shelley a half-share in my own mare Nanti, but it was not the same, for we could no longer ride together, and we watched her and Red Clover daily with an unspoken anxiety lest they too should fall victims.

Then, almost within days, Glen died of hard-pad distemper (after he had just recovered from a broken shoulder) and Shaun went into hospital where he remained for seven weeks with mysterious temperatures that defied all diagnosis to the last.

Duchess carried her calf to within three days of its full term, to our immense relief. Then suddenly she calved down one morning, when I was in bed with influenza, dropping her calf, unbelievably, through a hedge in Barn field and down on to the stony hollow lane ten feet below. It was the last blow that this terrible winter had for us, and it was bad enough. I remember hearing the shouts from the lane as I lay in bed; then pulling on my clothes and running with Bill to where Duchess bellowed at the hedge. We carried the calf home in a blanket, but there was nothing we could do. It died in the cowshed when it was an hour or two old. It was a fine heifer calf, full pedigree and worth at that time about forty or fifty pounds.

It was indeed a dreadful winter, and no one had the heart to think much about further improvements to the house. But all things come to an end, and the bad days passed. Spring came once more, with nightingales and new green grass; the butter kingcup yellow in the churn and the thick cream crinkling under the skimmer. Plum petals drifted on the rainwater in the tank where dead brown leaves had sailed before. All our young fruit trees were in blossom, a glorious sight, though every blossom was to be picked off by hand to let the young trees use their strength for growing. The garden was really a garden now and, best of all, Shaun had come home from hospital, well again.

Shelley, always a steady character, had taken the double blow of her losses with her own mixture of fortitude and quiet acceptance, but my good friend Mrs Williamson (who bred the gun-cat Vashti), hearing of Shelley's bereavements, sent her the gift of a blue-point Siamese kitten, Dromore Freyni. Siamese blue-points are muted versions of the better-known seal-points. Where Vashti is dark brown, Freyni is palm-willow grey; where Vashti's eyes are azure, Freyni's are china blue. These blue-points used to be considered very rare and were born here and there in litters unaccountably; but now that the factors governing heredity are better understood, breeders can have blue or seal to plan.

Catto was at first openly hostile towards Freyni, but later she became more subtle. She presented Freyni with a shrew. Shrews, though not actually poison to cats, disagree with them and make them ill and sick. Most country cats have shrew trouble when young and get over it, like measles; once eaten, shrews are never tried again. Catto would still catch them sometimes but now she never ate them, leaving them lying about on garden paths for me to see and admire. But Freyni was very young. She was so astonished when Vashti actually offered her something that she almost didn't eat it, out of sheer fright at such uncharacteristic behaviour. She ate it in the end, however, probably because Vashti willed her into it, sitting silently watching her until she had done so and then still hanging around hopefully to gloat over Freyni's discomfort.

Later in the summer, when our first male Siamese, Simba, arrived (Patrick having gone away to friends) they both offered him shrews. But he was nine months old at the time and must have got over shrew trouble in the spring before he came. He wasn't deceived.

The bathroom, almost our sole topic of family conversation except for milk yields and manures and carburettors and suchlike, was promised by Bill to begin on the first wet day of our second summer. We made a further trip to the Guildford shop and chose the bath, with basin and accessories. Would we have it fitted by the harvest? the hot, dusty, weary harvest? Oh, well before then, he hoped.

We did not have one wet day in the whole of that long, dry, tropical summer. The ground burned. Nobody thought about the bathroom and we carried water, water and more water to our newly planted fruit trees and still saw several die with brown and shrivelled leaves. Our first corn crop foundered under the double attack of drought and rabbits and we ploughed it back in June. Usually, with an ordinarily showery spring, the corn can grow above the rabbits, even on land so heavily infested as ours was. But this is hot dry land at the best of times, and unretentive of water. In that spring and summer only the established and deeply rooted things survived.

With plenty of surplus milk from Dawn and Duchess but no calf from either, we bought three calves to rear, and fed them with their gallon of milk each a day by bucket after the milking. They were called Gentian, Whinberry and Coronet. Bill increased the poultry and started a small nucleus of two geese and one gander. The gander, Evil Eye, is still unfortunately with us. No one but Bill can see one single thing in his favour. Without exception

Evil Eye is the most truculent, aggressive creature I have ever encountered. Either you carry a large stick or you run very fast indeed; if you do neither you will get quite shockingly pecked. Bill doesn't bother, he just grasps the bird round the neck, swings him round and round a few times like a salad-shaker and puts him down again. By this method Evil Eye can be rendered so mentally confused that further attacks are unlikely for at least five or six minutes; by which time Bill is elsewhere. No one but Bill has ever tried it; the gander is a very big, very overpowering bird, with a cold fish-blue eye. He was always able quite easily to drive the ponies away from their hay, or the cows from their water tank. (Nanti was terrified of him from the first sight; but she is easily bossed about. Red Clover, half her size, has bullied her for all their life together.) The cats cunningly avoid any contact with him, their natural sense of dignity keeping them clear of probable brawls.

Freyni soon became a passionate hunter and Vashti, seeing someone else being efficient in her own line, hunted less. Instead she followed me the more. She gardened when I gardened, walked when I did, went shooting, riding, and milking when I did. Most of my writing I did with my loose-leaf folio awkwardly balanced on my chair-arm because Catto was sprawling on my knee. I bathed with her sitting on the flat edge of the bath, and finally gave in on the last point, letting her sleep on my bed. This crowning concession so went to Vashti's head that she began to rush upstairs earlier and earlier each day, in case I might change my mind and put her back in the kitchen with Freyni and Simba, until finally she was going to bed at two o'clock in the afternoon. Then suddenly she must have thought how silly this was and gave it up all at once, going decorously upstairs with me at whatever hour I happened to finish the evening's writing.

Vashti adores all kinds of movement. She loves being wheeled in a wheelbarrow, no matter how rough the route. I have wheeled her up and down the stone steps at the garden gate, right down the drive and back, and round the farmyard. She will always jump in if she sees me trundling a wheelbarrow, so long as there is a little room somewhere in it for her among the stones, flower-pots, cement, or lawn-mowings, or whatever else I am transporting. She likes being swung, either lying limp across my two hands or sitting in a shopping basket or trug, peering happily over the side like a child in a swing-boat. Often she will jump into the empty basket, banging the handle with her muzzle until someone is persuaded to pick it up and swing her. She also much fancies being whirled about the floors on rugs and mats, even keeping her seat when bumped across the doorstep and all round the bird cherry on the little south lawn.

Freyni is quite a different sort of cat. (Cats, of course, like all creatures, vary just as much as humans do, as anyone will know who has lived for long enough with animals.) Freyni is smug. That is the one word which is the most expressive of her nature. She has a smug expression to begin with, her mouth turning slightly up at the corners, her china blue eyes slanting and squinty (we adore squints in Siamese cats and are sorry that fanciers are trying to breed them out), her features small and precise, including little ever-moving ears and little tail — two aspects not approved at all by the Siamese cat fancy. She accentuates her smugness by sitting about in lonely places, such as the middle of a room when everyone else is by the fire or window, the top of the court cupboard or the bottom of the stairs, looking straight ahead of her through half-shut eyes.

She can, of course, abandon her smugness if she wishes to, just as Vashti can forget her hostile feelings. Then they tear about the house together with their tails flung over sideways like jug handles, thundering up and down the stairs with an amazing noise for their small size.

My dear mother one day had a serious shock over this. She did not know that the cats were being fey just then, for she had been out of doors, and when she came into the sitting-room the place seemed empty enough — until suddenly she noticed the chair covers. They were, it seemed, possessed of devils, for they flapped and heaved, and bulged out here, then there, then flapped again as if the chairs were hoping to take flight. My poor mother clutched the door-post and looked again; and then she heard a rending snort such as only Catto can make, and suddenly there was a Siamese avalanche from underneath the nearest cover and my mother sat down to get over it in quietness.

We tried the experiment twice of keeping an entire male Siamese at the farm as husband to Freyni, but both these experiments failed. The first, Simba, was a seal-point of charming manners. His one drawback was retrieving Shelley's numerous cacti from their pots; an exploit both painful to him and grievous to Shelley, who is a connoisseur of cacti. He also doted on catmint to a quite unheard of degree. All our cats like catmint; I grow it specially for them and they roll in it and eat it and go to sleep in it, and the catmint still flourishes exceedingly. But Simba was fanatical about it. He could never walk down the path without casting himself into the hazy depths of it.

I forgot about this one evening when I picked some of the long blue sprays to go with Mrs Sinkins' pinks on the kitchen table. Simba and all the other cats were perfectly trained about the kitchen table. They know they can sit on the Aga, the Bendix, the dresser, and either of the two smaller tables, but never on the long kitchen table. It was here that I left my jug of pinks and catmint.

The next morning, coming into the kitchen, ruination and debauchery met my eyes. The brown Dicker jug was on its side at the edge of the table, the water on the floor; catmint and pinks were strewn with lavish abandon, crushed and broken now, upon the table; and in the midst rolled Simba, drunk, happy, and past all caring; not even collecting himself when I came in.

He had never been particularly strong, and he fell a victim to some obscure blood disorder which would not yield to penicillin. He left behind him one litter only, by Freyni, from which we kept the exquisite seal-point Purachatra Thistledown, belonging now to Shaun.

The second male, another seal-point and also called Simba, was a great success physically, being strong and upstanding and hearty. But he was a mighty hunter and once returned after four days' absence, thin and weak, with the raw weal of a rabbit-wire right around his middle. We had no wires out on our farm and our neighbours said they had none: either Simba had travelled far or else there were poachers working hereabouts. We pulled him round from that set-back, dressing the weal and feeding him with cream and other extras. He had just about completely recovered when he vanished altogether from our lives. We hunted everywhere for him, calling him round the countryside for days, phoning neighbours and the police, but there was no clue as to his fate.

People told us that this was what always happened to those who tried to keep a stud cat as a pet. The Siamese cat books said the same: "You cannot keep the entire male Siamese loose about the house."

We decided not to try again, but to send our cats to studs at local catteries as did most other breeders. The tale of our changed policy, in favour of one Dromore Cheetah, must come later in this story.

CHAPTER ELEVEN

Baths, Blossom & Plumbing

Suddenly the weather broke and it was raining.

"I think," said Bill, "I might begin on the bathroom."

The same day, back came the old familiar mess into the house; the sacks of cement and sand, the boxes of tools, the planks and ladders and ropes. Bill's plan for the bathroom was to join together the rotting old priest-hole with the cavernous chimney space belonging to the fallen old wing, thus making a spacious room about twelve feet square; and all "out of nothing" as one might say, for this was hole-and-corner space, unused for generations.

The priest-hole floor was rotten (I should say that we don't think it ever was a genuine priest-hole, but rumour has it so); the chimney, of course, had no floor, and it was filled with rubble from the ruins. All this rubble, said Bill, must come out; and since we had to have a window in the bathroom (besides the tiny mullioned one already there), what could be more simple than to make the hole for the window and to use it for throwing out rubble and hauling up builder's supplies? The rubble naturally came out first, making a stark little mountain on the nettle-covered site of the old wing. Bill fixed a pulley system to the gable, employing me as builder's mate to fill the buckets for him to haul up. I found this very tedious when there was so much else to be done, but it did save trails of mess through the house and up the stairs.

All the materials went up on this supply-line for the reinforced concrete floor which Bill, amazingly, constructed all around him as a base for our primrose-coloured tiles. I have never understood properly what supported him, but every now and then he would appear at the window-opening and send down a dangling empty bucket for me to fill from the barrow on the path. Once or twice I sent Catto up, to her immense delight, and she would hasten down the stairs and through the house again, hurrying out to have another go.

The joists which Bill used to support his reinforced concrete floor were from ash trees he had himself felled on the farm and trimmed to shape on the saw-bench. The new window had, unfortunately, to be just where the outline of the old wing fireplace showed in the wall. Consequently, we have lost all trace of this historic feature, except in photographs; but it was unavoidable. Bill's delicate feather brickwork under the window helped a great deal to make up for the loss, though, historically, a loss like that cannot be compensated. But the alternatives of a dark bathroom or no bathroom at all were neither practical nor desirable.

A new window (complete with little holes all round the frame) was made in Guildford and delivered to the farm. Wall board, ceiling board, and a large boiler with cistern and innumerable pipes were bought. The boiler was fitted into the ground floor of the old wing chimney space, directly under the bath. The clearing of this space — filled up entirely with rubble — was another really tough job for Bill. But it was, again, space that had not been used for generations, that was wholly wasted, and was admirably suited to our purpose. So much rubble had now been trundled out into the farmyard that Bill had ample supplies for a base for a concrete yard, if ever we aspired to such grandeur. There was the rubble from the old chimney (inside and out), from three new window spaces, three fireplaces and the old kitchen floor.

The old wing inglenook, when cleared, made a perfectly adequate little boiler-room, with the floor excavated a foot or so to allow the boiler to be lower than the lowest (future) radiators. For the outer wall, the smallest leaded window obtainable was ordered and fitted — and this, I am happy to say, without in any way affecting the exterior outline of the inglenook, whose brick arch still looks out through the winter jasmine on the wall.

The chimney for the boiler presented a problem because the original chimney had, of course, been demolished above the roof and used for bathroom and boiler-room space below it. A new chimney would have to be built, but we did not want to take it up outside the house because of spoiling the timbered gable.

"If we build it inside the bathroom," said Bill, "what shall we do to disguise it?"

After a great deal of discussion and plan drawing we decided to take the chimney from the boiler-room through the bathroom floor and up the inside of the wall as a semi-circular brick flue, later to be faced with cement and painted white with the bathroom walls. The base of this column, which widened to allow a bend in the flue, was happily disguised by a semi-circular seat. This Bill made by building a cement-faced wall, crescent shaped and about eighteen inches high, a good seat-width out from the chimney; the seat itself was made from ash wood, felled, sawn and finished on the farm by Bill himself.

In no time at all, it seemed, once the initial rush was over, the bathroom had materialized. We were invited by Bill to dig out from obscure drawers all the presents of bath-cubes and back-scrubbers and towels and sponges we had so hopefully made to each other on past Christmases and birthdays, and enter into our kingdom.

We could literally *bath.*

Did it matter — need anyone notice the fact, let alone mention it — that there was no lock — nor indeed any fastening whatever — on the four-foot high door? That there were no curtains yet at either window, no tiles on the floor, no airing cupboard round the gargling cistern and that sacks of cement and boxes full of spanners and suchlike cluttered up the floor space?

We could *bath.* And such a noble boiler it was that we could bath interminably, four or five in a queue, as deep as we liked and no need for any to call out, "Leave some hot for me!"

Vashti adored the bath from the very first sight. Sitting on the edge when I was in it she would lean down, patting the steaming water and licking her paw. Hot water, especially scented hot water, is one of her favourite drinks (we always fill the cats' water dish from the kettle or hot tap) and when I have the bath very deep she will lean low enough to reach it with her tongue, lapping ecstatically and teetering horribly on the edge. Once she slipped and fell in with me, and I don't know which of us suffered most from shock: a cat is not the most comfortable thing one can think of to have in one's bath.

Going into the bathroom, one can often find her lying in the damp empty bath, her dark points looking dramatically beautiful against the shining porcelain. I remember one awful day when she tore upstairs in a transport of joy and cast herself as usual into the bath, only to find it full of cold water and seashells, together with Shaun's submarine and tug-boat. She came out like a depth-charge, furious and soaked and flipping water everywhere. She wouldn't speak to anyone for at least an hour, not even to me: but it didn't cure her of her fancy for the bath; it only induced a preliminary cautious look instead of the old abandoned headlong leap.

Gradually the bathroom began to look less like a builder's workshop. The primrose tiles went down, the heated towel-rail and the basin were fitted, and then Bill began to make the spacious airing cupboard, using wood from a large wardrobe he had bought at a recent sale.

Mother and I were making curtains from some French gingham I had bought for summer dresses, Bill fitted mirrors to the walls and Mrs Fisher made us an "opening presentation" of a large luxurious bath-mat. Nobody thought about the door while these more urgent matters occupied our time; but there was not much wrong with it, according to our reasoning. It was so low that you couldn't possibly hit your head on the lintel, which comes about to my chest, and a fastening was not considered absolutely indispensible since we could always splash and sing.

Much later, Bill fixed a wooden latch which can shut people out, but there is still no system for closing the door from the outside. Whenever we want to shut a "calling" Siamese queen in there we have to devise complicated schemes with bits of string.

This winter we began to face the fact that Duchess and Dawn were both sterile, presumably as a result of the *Brucella abortus.* Months passed and neither of them could be got in calf, until finally both were sold as barreners at a tenth of the price we had given for them. We retained the three heifer calves, now nearly yearlings, and later got them in calf without trouble. Now we were buying milk again, and there was no more cream and no farm butter for the family.

In some ways I was glad of the breathing space. I had been rising at six in the morning to do the milking for nearly two years, seven days a week without a break. Somehow I had never seen to it that others learned to milk. And this winter Mother decided to set up house with my sister in Kent and I was managing alone again. The milking was one thing that I need not do for a while, and there was no butter to make and no dairy utensils to wash and sterilize each day.

There was, however, a new job which was to be mine each winter now, and that was the pruning of all the young trees in our two orchards, a total of well over three hundred. In this matter all praise should be heaped on the advisory department of the Surrey A.E.C., who sent their horticultural experts out to us and guided all our work.

On most fine days in the winter I would muffle myself up in coat and scarf and balaclava helmet, going up to Yew Tree Orchard or the small domestic Little Orchard with my knife, well-sharpened, and a pair of secateurs. And though she hates cold as any tropical animal does, Vashti would usually come too: but often she would hurry back to the house again, grumbling loudly, after following me down half a row or so and chatting to me as I worked.

Pruning is a delightfully soothing job, slow and satisfying. Each tree has a separate personality, to be considered by itself and for itself alone. There is no universal rule of pruning to be applied to each tree in its turn; it is this that lifts pruning to a place among the arts: for it is an art to prune a young tree so that the most perfect shape and balance, the right amount of growth in the right places, are induced.

More delightful still is the job of blossom-picking in the heady month of May; especially when it comes after arduous weeks of hoeing six-foot circles round the trees. For two or three years after planting, the blossom is all hand-picked in springtime, so that strong growth can be made and the trees properly established. Forgetting for a moment the days when I have picked blossom in a snapping north-easter, shivering in my heavy jacket with the collar turned up, it is pleasant to remember the soft days when the nightingales sang all around me through the long afternoon and the sun was warm on my bare arms and bended neck. Strange, the widely held belief that nightingales are silent through the day. I suppose their name is partly responsible for this, and the drowning of their daytime voice by all the singing birds of spring. But the voice is there for the quiet discerning listener. Only at the somnolent hour of noon or on days of cold wet wind is it absolutely silent.

Some still debate the merits of the blackbird's song compared with that of the nightingale. I think that the two are not really comparable at all. Each is good in his own line; but the blackbird is the errand boy and the nightingale the maestro. The blackbird has a tune, often a good tune and almost always original, quite distinct from other blackbirds' tunes, and he whistles it well and clearly. But it is only a tune. The nightingale, being a true musician, first takes a tune and then develops it, embroidering and enhancing it, introducing other tunes and phrases. He is in the great tradition. (It is, of course, only the male bird that sings, however much the old song may insist, *as she sings in the valley below*.)

There are many nightingales all round Punch Bowl Farm; I have heard them at all times and places. But when I remember the nightingale's song it is of the blossom-picking that I think, and the falling flowers and the warm wind and Vashti rolling in the petal-drifts. On such days one is veiled about with drifting petals — clouds of shell-pink petals lifting and falling in the light air. Vashti, drunk with spring, leaps at the floating showers, catching them in her paws and playing with them. Coming down to the farmhouse from hours of this dreamy petal-enchantment one is accosted by members of the household who pause to pick off blossoms caught and held in one's hair and on one's shoulders.

This is the kind of job that is conducive to thinking about the book at that time being written. Most writers, I suppose, are apt to drift into such contemplations while doing things unconnected with writing. I know I do myself; often with catastrophic results, as when I am cooking or shopping or doing any other kind of job that is not wholly automatic. I know, for example, that I must never, *never* think of books when I am pruning, which is a job calling for one's whole intelligence the whole of the time; but blossom-picking is the same with any tree, one's fingers moving of their own accord among the bright branches, leaving the senses to be drugged with blossom-scent and colour; the mind standing apart occupied with plots and situations and characters.

I once devised a scheme for reminding myself that I had put something in the oven. It was a tower of assorted kitchen items, such as a casserole, a jug, plate, cup and saucer, all built upon the kitchen table and the whole surmounted by the kitchen clock. Whenever I look at that, I thought to myself, I am sure to wonder why I put it there, and then I shall remember that I've put a fruit cake in the oven. But my mother is a very tidy person, and, seeing my tower, she would say to herself, "Why on earth has Monica left all those things there?" and tidy them away. Once we found a dish of sausages that

I had put into the oven for supper two evenings before. They were jet black and very shiny, like patent leather. They looked like a wonderful new synthetic material for making handbags.

It was about this time that Bill started considering central heating. He had always had the idea since buying the enormous boiler, and now there was the summer ahead of him for the work. Naturally, being Bill, it had not occurred to him to hire a plumber. He and his brother Sam spent hours with books on plumbing, with plans and blueprints and circuits. They went off in a van, returning with eleven radiators and a very great quantity of piping. They walked round the house tapping walls and floors and ceilings. Then suddenly they began.

In the garden outside the back door they set up a bench with a kind of lathe for making "threads" on pipes. A peculiar pile of metal shavings, like a robot's curly whiskers, grew underneath, and beside this grew a most impressive pile of pipes. The old familiar mess now suddenly reappeared about the house. I knew it so well. The spanners, the sacks and boxes, the drills and monkey wrenches, the snail-trails of unidentifiable substance on the stairs.

One might be in any room at any time when suddenly the wall or floor would erupt at some place admitting a man's hand, to be followed by a length of well-greased piping. The radiators were brought in and dealt around the house; oh, no! they were not connected yet, we must not lean on them, or sit on them, in fact we had much better not touch them at all.

But the day came.

Again I was gripped by the old impulse to go in or out of the house very quickly because of something Bill was doing. This time, of course, I went out, because indoors Bill was turning on the central heating, having got the water in the boiler very hot for the occasion. At the worst I expected the house to blow up. At the very least I feared volcanic floods.

Now, said Bill through the doorway, he was just going to turn on to see if there was an air-lock.

And what would happen, I asked him nervously, if there should chance to be an air-lock? But Bill had vanished from the door. There was a long, total silence. Then Bill reappeared. "Queer thing," he said, "I turned it on and nothing happened at all. No water, no air-lock, no anything."

"Well, really," I said, relieved, "what could you expect? Plumbing is a skilled job after all; you could hardly hope to succeed at such difficult and complicated work without the years of proper training usual."

"What on earth are you talking about?" asked Bill, a little crossly. "The thing's working beautifully now; you could toast on all the radiators; come and see for yourself if you don't believe me. But what I would like to know is, who turned off the hot-water valve without my knowing?"

We never discovered the answer to that.

In the spring I sent Nanti away to an Arab stallion for a few weeks. The heifers were all in calf and Freyni and Thistledown were in kitten to an unknown suitor, since we had somehow failed to get them away to the stud.

It was hard for Vashti, the maiden aunt, who somehow didn't care for family life and all that kind of thing. Kittens, she felt, were a great mistake: if she could do without them, why couldn't everyone else? The fact was that Catto soon became a psychological case over kittens, and if she had been human I would probably have had her analysed. Kittens were about the only topic on which Vashti would hold forth at length without anyone asking her opinion. She had only to be in the same room with a maternity basket to start spitting and howling her annoyance and distress.

Fortunately, as the kittens grew, she became a little more tolerant, or perhaps I should say a little less openly hostile. I have seen her jump into a chair where older kittens were curled up and vigorously start washing them, but growling all the while to show her disapproval. In the face of this terrifying experience the wretched kittens would lean apprehensively away from her, their small ears laid back nervously, until the whole domestic scene wound up in a grand slapping and admonition all round, kittens scattering like skittles, leaving Vashti sitting alone, snorting and furiously squinting.

Freyni's cross-bred kittens were sometimes delightful blue-greys having very close velvety coats, like moles' coats but with a silver light shot through them. They used to hide themselves, as kittens do, in all kinds of unthinkable places and were once lost for more than five hours before I found them asleep in a far corner under Shaun's bed. One day Shelley went into the kitchen and found three kittens only in the basket. The other two, she said, would be sure to turn up somewhere: and she went to pick up the empty cream-pan which had slipped from where it was propped against the wall and fallen, like a Chinese coolie's hat, upside down upon the floor. Underneath, blinking in the sudden daylight, were two small kittens together with a ping-pong ball.

CHAPTER TWELVE

The Townsman in the Country

We never heard the telephone bell in our first years without half expecting someone to tell us, "Did you know that your heifers are out?" or, "Your ponies have just gone past our gate towards the Punch Bowl."

The hedges of Punch Bowl Farm had for many years grown almost wild; they were too lush at the top and too thin at the bottom, which is the way of all unpleached hedges. Persevering animals can walk right through them, though to the casual observer there is no visible gap. Cows are, in this respect, a great deal worse than horses, we found, but horses are quite bad enough. When, in addition to weak hedges, one had to accept the risk of wandering trespassers leaving gates ajar, the chances of having to drop everything and go out in search of animals were much increased.

It seems to me that the old mutual antagonism between townsman and countryman will never be much improved while townsfolk treat the farmlands as they so often do. This is old stuff, of course — the way the townsman likes to think of the country as his playground, of the grazing land as his picnic spot and the copse as a dump for his ice-cream cartons and empty bottles. And, on the other hand, the way the farmers shout aggressively at innocent hikers merely walking through his fields. (Did they have a dog? Was it running loose? And if so were there in-calf cattle or possibly sheep in the fields? Dead calves and lambs are too often the result of what, to a dog, was merely a game. Did the first man through the gate make sure that the last one shut it? Or if they climbed the gate did they all go over at the hinge end to lessen the strain?)

Some townsfolk might be surprised at the way they seem to us here in the country. The town-born idea of clod-headed yokel and sensible townsman is reversed when the townsman is away from his own ground. Left-open gates are indeed a serious hazard, when two of one's neighbours keep bulls, and some are attested and some not, and no one has time to spend beating round the Punch Bowl for lost cattle; but gates are not the only hazard. I have seen a band of young men leaping into our Old Orchard hedge, stamping it down to make a gap for their womenfolk to walk through.

Consider the townsman's attitude to grass. Our new neighbour at Upper Highfield once accosted a party walking through one of his fields which was shut up for hay.

"We didn't know it would matter," they explained, "as it is only grass."

"But grass is a *crop*," said Mr Abels, "just as much as corn or roots, and quite as valuable. This field was *sown* to grass."

But they always thought that grass just grew, they said, and in any case they didn't know it was private land.

"Not private land?" said our neighbour, "and what do you think those are?" pointing to his cattle over the fence, "wild animals?"

These country attitudes may seem selfish and unfair to the town dweller. The whole trouble is that the sensible townsman suffers for the sins of the foolish. Since some trespassers have left his gates open and made gaps in his hedges, the farmer naturally suspects all, so that all may be ordered off his land. One has only to spend a wet night tramping through the rain in search of straying cattle, as we have done, to understand the justice of this attitude. We have hunted all the wild woods and heathland round here many times, by night and day, for wandering stock; and though sometimes the fault was in our own weak hedges, it was more often due to the passing stranger.

The particular night I mentioned, when it was raining hard and continuously, we were looking for the heifers Gentian, Coronet and Whinberry. We didn't find them; the rain and the darkness were against us, and also the fact that tracks were almost invisible in the torchlight and under the effacing downpour. But in the morning, setting out again, we soon began to realize that several times we must have been quite near the heifers in the night. We found them on the fringe of the wild valley and drove them back; and that day we fitted all three of them with collars and bells.

"Now," said Bill, "if the same thing happens another night we shall hear them. And the bells could be useful by daylight, too, in the valley and Punch Bowl where the undergrowth is so thick."

The heifers' bells on a summer afternoon make one of the loveliest sounds in a lifetime. On different notes, ringing haphazard as the heifers peacefully grazed together, the bells were a part of the dreamy solitude of these old pastures. As the summer slowly passed, the bell-music became such a part of it that I could never think of hot drowsy noons, leaf canopies, full-sailed white clouds and honey scents of drying hay without hearing in my mind the far sound of the heifers' swinging bells.

With the reassurance of the sound one could work at peace, picking blossoms, writing, gardening, pruning, and knowing that the heifers were safely within bounds, while the triple tinkle came and went upon the wind.

In time, one bell and then another fell off and was lost somewhere in the fields or hedge-bottoms. Bill had another idea. We might not be able to stop the casual gate-opening tripper, but we could do our best to prevent more escapes through weak hedges. There was not time enough in Bill's life or mine to layer and mend all our hedges properly. We had spent weeks on them at one time or another, but there was always the possibility of a thin place being discovered and exploited, especially by Coronet, the boss-cow of the three. She would go through the most unlikely places, making a wider gap for the other two behind her. Then away they would all be gone (and probably the ponies too) and away we would have to go after them, whatever urgent matter had been engaging us.

Bill's new idea involved collars as before, but poles were to be slung instead of bells. They would be more efficient than the bells had been, he said, because the poles were designed to stop the heifers breaking through the hedges, whereas the bells had only helped us to find them after they had broken through.

"But the bells were much nicer," I said wistfully; "and what about the times when people leave the gates open? The poles won't help us then."

"They might," said Bill. "They'll stop the heifers from getting away so far. You see, the idea is that we cut a pole for each heifer, about three feet long, and make notches at each end to take a rope tied round; and then we sling the pole round the heifer's neck with the rope. The pole hangs broadside across the heifer's chest and gets in the way whenever she tries to push herself through a gap."

The poles were so efficient that after a time we were able to abandon them completely, the heifers having become convinced that our hedges were now all stock-proof. And, in any case, when they were heavy in calf they settled down more staidly and lost the urge to roam.

The ponies rarely broke through hedges but they also rarely missed any left-open gate, leading either off the farm or into the garden. One summer evening just at dusk I heard a drumming of hoofs on the high track that runs beside the yew tree. Dropping my pen I rushed up the steep path from the farmhouse just in time to see Nanti vanish into Lower Six Acres, which was then empty of stock and standing open. Turning, I saw the gate into her own field wide ajar. It was getting dark and I was busy; I decided to leave her where she was until the morning. In any case she would not be easy to catch while she was excited just after her escape. Shutting the two open gates to prevent the other animals following her I went down to the house and finished the chapter I had been writing. And after that, over a supper drink, I mentioned the matter to Bill.

"I suppose you know the barway at the far end of Lower Six is down?" he asked calmly. "As there were no stock in there I left it down for the tractor."

There is no answer to this sort of remark.

It was now totally dark and Nanti was probably already in the Punch Bowl; there was nothing to prevent her.

"Perhaps she'll stay where she is," said Bill comfortingly. "The grazing's much better in there than in her own field."

Remembering hopeless nights of heifer hunting I resisted the impulse to dash out into the darkness with torch and halter and went to bed. But in the morning I got up very early indeed (surprising Vashti who is usually the first to wake, rousing me by gentle pattings of my eyelids and loud purrings while sitting heavily on my chest). It was late summer and Shelley was on holiday in a Bavarian farmhouse at the time and so was not available to help me. I thought to myself, this will be a bit of news for my letter to her.

Nanti, of course, was not in Lower Six Acres. I stood on the corner of the field for a moment looking at the thick green aftermath and remembering our first furrows, in this field, with the shilling horse-plough. Then suddenly I saw a red movement, and two more; it was a vixen with her young cubs trotting lightly in the dewy grass beside the hedge. They had not seen me, so standing still I was able to watch them till they moved from the field some minutes later. The vixen was doing her best to instruct her children in the arts of hunting, but they, childlike, did not care to be instructed. It was, after all, a perfect summer's morning: the grasses were heavy with the dew, and the rising sun made all the valley sparkle. The cubs wanted to play, and play they did. The vixen trotted decorously all the way up the hedgeside, sniffing and snuffling, pausing and peering, looking at her children disapprovingly and waving her brush. Her two cubs leaped and rolled and frolicked behind her, yapping at each other — worst possible crime in a hunter — as they did so. Not until they had left the field at the further corner did I move from my place and head towards the open barway.

From this place Nanti's direction was unknown and I had only her faint tracks for guidance; faint because days without rain had left the ground hard and dry. I began to discover how many clues a straying horse can leave for the tracker, even in conditions like these. With no serious difficulty I followed her trail for several miles, through the Devil's Punch Bowl, across the Portsmouth road and over Hindhead Common to Haslemere, where I lost it on the metalled road. At that time of day the dew was a great advantage, for it showed up the dark path through the grasses where Nanti's hoofs had brushed the silver from the green. Dust too was an ally, for it showed here and there a place where her foot had scuffled it. At Small Brook in the heart of the deep valley I saw the place where she had drunk; but from here the tracking was more difficult. Several times I had to retrace my way from a false trail made probably by deer or a ridden horse the day before, the marks so slight that it was hard to be sure whether anything had really passed that way at all. The kicked stone, the snatched tussock of grass, the broken flower; all helped to show the way the mare had gone. But at the tarmac road the last mark I was to find showed where she had stepped off the common.

Which way? I walked up and down the road, hungry, thirsty and tired by now, and my dusty sandalled feet ached. No one had seen or heard a horse passing down the road. Road sweepers with their barrows had not seen her. I made inquiries at a nearby house and found the family at breakfast. Here I was offered a welcome cup of tea, and the use of the telephone to ring both Bill and the police. Bill said he would come out to meet me on the Portsmouth road; he had thought I was still in my room. The police had no news. I set out across the common, tired as at the end of a long day, eating the brown crust I had brought to help catch Nanti.

Later in the day the police telephoned from Hindhead; a chestnut mare answering to my description had been found near a Haslemere farm and turned into a paddock. I phoned the farm, thanking them. She was a sweet mare, they said, and no trouble, but grazing was short at this time of the year. I knew it, I said, and would fetch her back at once. . . .

Do the leavers-open of gates ever think of things like this, I wonder? It would have taken the trespasser half a minute to shut the paddock gate behind him; but I lost the whole of one day in getting Nanti back, as other days had been lost before and as undoubtedly they will be lost again.

People who love the country enough to read books about it are not likely to be among those who harm it, but because they also will be suspected (people half a field away look very much alike, even if one is a first-rate judge of character, which I am not) it may not come amiss to put the farmer's case. A recent speech made by Lord Mancroft in the House of Lords described one farmer's encounter with a picnic party on his land:

"They laughed off cheerfully his remarks about trespass, and invited him to call on them the next time he was passing through Wandsworth. To their surprise he did. He was carrying a large sack. Their surprise was all the greater when he proceeded to pour out on to the sitting-room floor all the empty beer bottles, empty tins and orange peel they had left on his property."

These people, Lord Mancroft added, were typical thoughtless litter-louts, who were in the great majority. There also existed, he said, a staggering number of deliberate vandals.

"I have case after case," he said, tapping his portfolio, "extraordinary cases, of people who have gone into farms, taken every gate off its hinges and dumped them into a pile, just for fun. There are people who have gone into an agricultural implements park on a Sunday and have slashed every tyre; and people who have deliberately let out cattle, mixed up attested and non-attested herds or let them get out on to the highway.

"I do not know what we can do about it. I am depressed to find that the education of the public in the way to behave in the country has not been the success we hoped. The Government is anxious that the utmost publicity should be given to this problem of country good manners."

I hope I may perhaps have helped a little with this quotation and with the tale of our own and our neighbours' experiences. In fairness I should add that we ourselves have not found the downright vandals very numerous, but the damage they do and the trouble they cause make their numbers seem the greater.

Litter-louts, however, certainly are in a great majority, as Lord Mancroft rightly says. Usually, I think, they are townspeople, though by no means always, and they cast their litter about wherever they are, in town or country indiscriminately. It was shocking but not surprising to hear of the seas of litter left in London's streets and parks by Coronation spectators in June, 1953 — twenty tons in the streets of the route alone. Surely, if full bags and cartons and bottles could be carried to the City, they might reasonably have been folded and packed when empty and carried home again? And consider the mind of a people who, while appreciating the beauty of their decorated Coronation route, could at the same time spoil it with their refuse. This is the same mind that, coming to see the glory of gladed bluebells or the hills of August heather, sullies them with rubbish from picnic meals, with whole newspapers and jaded cardboard packing boxes.

There are two things that I think might usefully be

fitted into every schoolchild's curriculum. One of these is swimming — the deaths from drowning every year are often pathetically needless — and the other is a true appreciation of our land. It is a small land, terribly overcrowded, and its diminishing beauty is largely in the hands of its uncaring natives. The few who do care are like a small voice crying in the wilderness.

CHAPTER THIRTEEN

Calves & Bees

During our fourth year at the farm I was twice in hospital, first for observation and the second time for a major operation. All sympathy for me was wholly wasted, for I freely admitted that I enjoyed almost every minute. There was no cooking to do, and no housework; no writing, because all attempts in this line were quickly frustrated by nurses with thermometers, nurses with meals, nurses with wash-basins and other hardware, nurses with dressings and doctors with shy bedside manners (it was a famous teaching hospital; I recall once being the *first patient* of one very young man; we spent the whole time discussing what his father should do in his retirement).

I found my fellow patients interesting, the meals quite good, the nurses charming, and the drama of an ordinary hospital day sufficient to ward off all boredom.

There was, for example, the high-born old lady in a private but by no means soundproof cubicle leading off the general ward. Being under the impression that she was in a first-class hotel she naturally found some fault with the cuisine, the staff, and the general management as a whole. We would listen, enthralled, to her demands to see the proprietor, her lecturing of the "maids" and her shocked accusations that her clothes were being hidden. I know that it was sad, of course, but even sad things have their lighter aspects and on the whole I think she was quite happy.

There was (during my summer stay, when beds were wheeled out into the garden) the fifteen-year-old cockney girl, sitting up glamorously in bed with a bubble-cut and purple lipstick, flapping her lashes at the stokers going by with their buckets and shovels. There was the young man under the trees who spent his time studying the sky with obscure instruments; the negro with an Oxford accent; the woman whose visiting son always spent the whole hour listening to the football through her headphones. There was the drama of the young and good-looking sister-in-charge and the doctor who (we fancied) obviously adored her. I think I may say that this romance was the daily highlight of our entire ward, looked forward to by all with as much warm interest as the Next Instalment of Our Serial. The time was usually nine p.m., just after Lights Out. A soft green light was focused over the sister's table and the stars of our play would come in; she so lovely, he so shy and handsome. Nothing ever happened, I make haste to add. Nothing, that is, but the small glance, the brief silence, the smile over the registers and papers. It was enough for the ward. In the shadows round the walls every head was propped on pillows lately flattened for the night, one woman going so far, I noticed, as to put on her long-range glasses as the quiet footfalls came along the passage.

These were small things, but they made up our day.

Two aspects of my illness that I regretted were; that I missed the full delight of Nanti's newborn colt (a fine chestnut, just like his sire) and that I had to give up, from now on, a great part of the farmwork that I had so much enjoyed.

We called the colt Sari, which is an Arabian word meaning Fleet of Foot, for he was of Arabian ancestry on both sides, his forbears the desert horses, Drinkers of the Wind. He was born in a rainstorm, early one May morning on the slope of Hanger field, and he was cantering around on that same morning before I saw him. He was brought past my bedroom window with his mother before breakfast so that I could see how big and strong and beautiful he was.

When he was three weeks old he had a day-long engagement with the staff photographer of a famous magazine. He and his mother had the camera around them throughout one sunshiny day. I missed that day, because then I was in hospital, but I heard all about the one hundred and fifty photographs, some of them coloured, from which only eight would be chosen; and about how Sari had behaved — so proud and yet so babyish, lying in the lush grass with a garland of poppies placed round his neck, or following his mother, long-legged and dreamy-eyed.

The chosen photographs were published later on a "double spread" headed "To Nanti — a Son."

That same summer the three heifers all calved down. Two of the calves were heifers and one a bull. Short-handed as we were and myself being of little use just then, we sold Gentian and Whinberry and the bull calf, keeping the two heifer calves, Tiara and Serenade (daughters of Coronet and Gentian) to rear on Coronet's milk. One thing my illness had done for the farm was to spur others into learning to milk. Now Bill and Shelley could both milk, and Shaun was learning.

The two calves used to be tethered on one of the new lawns when they were old enough to enjoy the sunshine. Soon they began to graze a little too, and the lawn grass was short and palatable for them. We used to mow a boxful for them when wind or rain kept them in the calf-box.

Anyone taking a deck-chair into the garden, when the calves were there, had to be careful to set up the chair out of range of both tethering-chains. Calves all love licking things; they don't much mind what; your shoes, your bare legs, your chair or your book — all will do quite nicely. They all love leaping and racing and they are always much stronger than you think they are. Everyone in this family grew so used to seeing Shelley or me tearing furiously round the corner of the house on the end of a tethering-chain that the only comment would be, "I see Mother's putting Tiara out," or "Was that Shelley, just now, with Serenade?"

It was during this summer that we made the sort of mistake made by novices from time to time. It started with our thinking that perhaps we knew better than the generations of farmers who had always done the thing quite differently. The subject was calf rearing.

Why, we said, did people always feed them from a bucket? You had first to milk your cow, then take the milk up to the farmhouse and strain it, then boil a kettle of water and measure out into buckets the ration of milk and added water for each calf. You had to take it back to the sheds and hold both buckets while the calves drank their milk — if you didn't they would certainly butt it over — and then you had to take the empty buckets back to the house and wash, scrub and sterilize them. Why not, we said, just milk two teats for the house and turn the calves on to the other two? There would thus be less milking, no bucket-holding and no washing and sterilizing; and surely it was a much happier arrangement for both cow and calves, the one being able to lick and nuzzle her child and foster-child, the others enjoying the mothering and the natural feeding.

I watched them sometimes, all three so pleased and happy to be together, and thought how wise we had been and how foolish were all the hundreds and thousands of bucket-feeders. At one stroke we had saved ourselves work and made three other creatures happy.

Up to this point we had not made any irretrievable mistake. Many farms do have a "nurse cow" kept only for the suckling of young calves; but, after weaning at three months or so, such calves are never turned out to graze with milk cattle until they themselves calve down, at two to three years old.

Our real error was in turning our calves out with Coronet, for company, about five months after weaning. Calves have very long memories and ours at once began to suckle. We separated them again but Coronet, being alone, fretted for the heifers and broke through the hedges to reach them. We did not want to return the calves to the sheds when they were now old enough to be at grass, and our hedges were not proof against Coronet's determination.

Bill began to use his ingenuity again. First he planned a sort of super-outsize brassiere for Coronet's udder, but it never stayed there, she hadn't that kind of figure. We used to find it lying about trampled in the fields and all her milk gone, as usual. Next he tried spiked halters on the heifers. Basically, these are the rather superior leather head collars worn by the quality among cattle in the shows. On this elegant foundation one fixes a row of strong spikes along the nose-band. The spikes are three-inch galvanized nails. A band of canvas protects the cow's face from the nail-heads but the spiked ends stick out in front and, when the heifer noses towards an udder, they poke the udder first. The udder, of course, quickly moves out of range; at least it does in theory. But our heifers used abominable cunning. Serenade soon learned to push an exploring tongue up through the spikes, suck in a teat and carry on, without inflicting so much as a prick. Bill tried longer spikes, and curved spikes, and double rows of spikes, and spikes with canvas bands across their gaps where Serenade's tongue came up. But the heifers learned to bend the spikes sideways and to rip the canvas off. Soon the elegant expensive leather halters were as shabby as something dug up from a corner of the yard.

Bill tried a plain canvas flap hanging over their muzzles, with no spikes at all. The idea was that they could graze but not reach anything dangling down in front of them. We soon ran short of canvas, for these flaps were very impermanent. Also Tiara taught herself to blow hers out of the way.

We tried bitter aloes on Coronet's teats, but the heifers adored it. Nothing was really effective until Bill bought an electric fence and set it up between Coronet and the heifers. This was miraculous, especially in wet weather, and it also gave us a great deal of harmless pleasure when watching trespassers step over it.

It is much simpler, however, to feed your calves by bucket from the start, unless you know that your hedges are stock-proof. It is also a good plan never to try the obviously easiest method of doing a thing until you have first ascertained why thousands of intelligent farmers go another way about it.

Shortly after the electric fence was put up, a pair of town young women came walking down Barn field. The first thing they touched was the electric fence, and having been more or less pitched over it by shock they were unwilling to go back. Coming nervously onwards, the next thing they saw was Tiara, followed by Serenade, gallumphing up the field to see who it was and why, and whether they had anything nice to eat on them. With the heifers on one hand and the fence on the other, the young women were in somewhat of a dither, but as they were by now nearer to the farmyard gate than to anywhere else they came on, walking backwards and in circles.

Bill, in the yard, began with our usual opening gambit, "Have you lost your way?" and on being told that they didn't think so, asked politely, "Didn't you see the notice, 'Private, Farm Only'?" They didn't think so. Oh, well, said Bill, since they had come so far they might as well go on through the yard now and out through the Old Orchard to the lane; but perhaps they had better keep a look-out for the gander, who was very aggressive, and for the colt, who was teething and nipped hard if you let him. This was mere statement of fact, but perhaps a little unnerving to strangers.

We never saw that couple again.

"One day," said Bill, "I'll put another notice up in the top fields, 'Beware of the Apiary.' That'll keep the worst of them off."

Our bees began in the comic-sudden way of most of our ventures. Bill was mowing grass for silage in Lower Six Acres; it was lovely thick clovery grass that we had sown and grown ourselves, and we were as proud of it as anyone else might be of a bowl of first-prize roses. It was the kind of day when the sun is metallic on the green leaves and the air drunk with the smell of mown grasses: and it was the kind of weather when dogs pant in the shade, cattle drowse under trees and bees swarm. I don't know where the dogs and cattle were just then but the bees were suddenly swarming on Bill's tractor. One minute they were not and the next minute they were. Some people might not have had the presence of mind even to stop the tractor, leaving it to blunder on its way with rattling mower right over the far fields to the Punch Bowl. Bill stopped it, and then tore down to the house with his hair and shirt and trousers full of bees. For five minutes there was language in the bathroom that I would not care to repeat. Then suddenly Bill thought about that swarm; *a swarm of bees in June is worth a silver spoon.*

"I may as well have that swarm as anyone," he said, appearing with a face like apple pudding, and he tore back to Lower Six Acres. I will say for Bill that he has presence of mind, and, of course, a fair sense of values: he must have, or he wouldn't have hived that swarm; and then finished his mowing, with one eye nearly closed and a jaw like a heavyweight boxer's.

Nowadays we have five hives (just next to my writing hut; I cannot imagine why) and Bill thinks nothing at all of bee-stings any more. The other day we were working together at ridging potatoes. We were in a part of Upper Six Acres that was newly cleared from brush, and were just in front of six more hives belonging to a friend.

"The bees are a bit wild to-day," said Bill as I mounted the tractor. "Perhaps you'd better wear the bee-veil."

"What about you?" I asked, donning the large dirty straw hat with alluring black veil hanging round it.

"Oh me, I don't mind," said Bill, grasping the handles of the ridger. "Carry on."

At a pause for breath later in the day I sat considering the unfamiliar lines of Bill's countenance. Did it really suit him, I wondered, the heavier jaw-line, full face and cunning-looking eyes? He reminded me of the villain in a strong American film.

"I feel just like the Indiarubber Man," said Bill cheerfully. "One minute I'm out here, and the next minute I'm out here, and then my head goes up."

CHAPTER FOURTEEN

The Price of Solitude

Seekers after solitude in country places must almost always pay a price. Usually it is paid in educational and household difficulties of a kind not known in towns. For various reasons we did not leave Shaun at the village school for more than his first year, our main reason being the dangers of the blind corkscrew lane to the village. Lanes such as this can hold more danger for a young carefree child than can, say, Piccadilly. There are no pavements, no verges, no zebra crossings, no clear vision of the road ahead or behind; and, worst of all, what traffic there is is so infrequent and often so fast and silent, coasting downhill, that a child is caught unawares. The very fact that traffic is rare lulls a child into false security, into strolling dreamily along the middle of the lane or dashing from side to side for larger red campions, thicker primroses and lower hanging honeysuckle. And then, suddenly caught, there is no refuge; only the steep green banks walling-in the narrow traffic-way.

This meant that I had to take Shaun all the way to school and bring him home again; the length of the lane to be walked or ridden four times every day, for in those days Bill had the old car in Wales or some similarly remote place for weeks together. I had not the time to spare for this; so that we decided to send Shaun to school in Godalming until he was old enough for the next almost inevitable step for isolated families — boarding school.

This too had its snags, though we were blessed again by that good neighbourliness we had encountered so often since we came here. Transport to the new school was easily managed through the kindliness of our neighbours the Jupes across the valley at Ridgeway Farm. For three years Mr Jupe took Shaun in his car to Godalming every morning, so that the boy had only the short and beautiful walk down the bridle path, over Small Brook by the stepping-stones and up the wooded rocky way to Ridgeway Farm.

The new school was six miles away and the return journey was made by bus for five and a half miles, Shaun walking the last half mile up the stony track from the bus stop and through two of our own fields. Half a mile didn't seem much in summer weather when the fields were dry and full of flowers and singing birds; but it seemed very long to a six-year-old in winter on days when the sleet drove down and the dark came early and his hands were hurting with the sodden frosty cold.

Shelley too, though much older, was greatly handicapped by the weather. Her school, about three miles away in another direction, was not easily reached by car from our region except by driving all the way round by Haslemere, the direct track over the common being rough and very stony. For the years that she had Tarquin she rode him to school, for a year after that she walked, and then she bicycled. The weather was always an important factor in these long exposed journeys, and often heavy rain, hill mists or snow would prevent her from setting out at all. Absences due to weather all build up to make examinations more formidable. Also, after illness a longer convalescence is required to face a journey of six miles overland every day. Schooling was unavoidably interrupted to a degree that would not happen in a town.

But we have never thought that school was the most important thing in the life of a child. Our children lost on the purely educational side, but they gained immeasurably in country living, clean air and farm food; and they acquired a sense of reliability by handling farm stock and implements, and by being needed and valued for haying and harvest and similar farm crises. They were part of the farm, and they accepted responsibility in a way town children rarely can, for lack of opportunity.

When Shaun was nearly eight we had to decide whether we were going to take the risk of his probably failing the County Entrance Examination (and at an age when no good private school would be likely to have a place for him), or whether it would be better to establish him now, while he was young, at a suitable boarding preparatory school. From all I knew of my family we could not rely on his gaining a grammar school place. We are all (I except Bill) one-sided, rarely being of much use in mathematics. I never passed an arithmetic examination in my life. If, in my day, an Education Act had made proficiency in arithmetic an essential for entry to grammar schools, I would certainly have been one who, as they so oddly put it, would not benefit by further education. And yet, now, I can earn a decent living as an author.

Shelley in her turn was rejected; and yet she had just previously tied with one other child for first place in an inter-schools essay competition. I suppose her mathematics lost her a place, but do the examiners never consider that we are not all exactly balanced on all sides? Sometimes a child who is weak in one subject is twice as strong in another. (Which leads one to wonder, following the theme to its logical limit; was Mozart any good at maths.? Was Newton good at English literature? Or did Shakespeare shine at science?)

We decided not to run the risk, and Shaun was "put down" for a suitable public school with its own preparatory house. We should have liked to have him at home, with a good day school on which to rely, but with the present educational system and our own isolation there seemed to us no other way out.

Against these educational problems my household difficulties may seem slight; but they probably accounted for my illness and subsequent periods in hospital. One cannot, of course, expect a daily help to come all the way up our steep lane from the village when there are jobs in plenty waiting near at hand. In six years at the farm, I have had just eighteen mornings' daily help. And that, I think, was a kindness following my operation. Now I have no help of any kind in house or garden. And the writing must be fitted in somehow, as well as such farm work as I still do.

In such a situation one is not always entirely helpless. Labour-saving devices make a great difference, and when we had the main water and electricity I was able to install a Hoover and a Bendix washer, the two implements that save me more work than any others, except perhaps the Aga cooker. I soon cut out all floor polishing by using the mop-soaked-in-oil system, and I lowered all my standards of household cleanliness and tidiness. It was the only thing I could do. Instead of cleaning all rooms every day, I now do only the kitchen daily and let the others take their turn.

Once I carried labour-saving a step too far. It was when I was churning the cream for butter, and for some reason the butter would not come. The cream had "gone to sleep" as country people say; it was probably too cold, because for butter-making cream should be at about sixty degrees Farenheit.

Now some washing machines have arrangements to take a churn and to do the work for you, but the Bendix hasn't; its machinery is too complicated and hidden away. Why not, I thought, put the churn *inside* the Bendix, well padded round with cushions? At three hundred and thirty-three revolutions a minute it ought to make the butter in a very few minutes. Remembering the small circle of ventilation holes in the churn lid I stuck some adhesive plaster over them, then carefully committed the churn (a gallon glass one containing half a gallon of cream) to a Bendixful of cushions. When all was firm and safely packed and I had seen that the water was turned off, I shut the porthole door and switched on.

For a few minutes the cushions spun round looking like a kaleidoscope in the porthole window, and then

I switched off to look inside. Extracting the churn, I stared at it disbelievingly. There was no cream in it *at all.* Not a drop. Where had it gone? I pulled out all the cushions, expecting to see them smothered in thick cream, but they were as clean as when they went in. I looked in the Bendix and that too was dry and spotless. It was the Case of the Vanishing Cream. Perhaps it was all down in the bottom somewhere, I thought, below the clothes drum. Perhaps if I were to put a clean basin to catch it and then switch to pumping out I might save it. This I did, and sure enough out poured the cream. I should think I would be safe in asserting that never before or since had anyone seen half a gallon of cream come out of the pump-pipe of a washing machine. Collecting it all (it was looking rather frothy now) I put it back inside the churn; half a gallon of cream must simply not be wasted, wherever it may have been travelling. After all, one washes the butter granules seven or eight times after they have "come," so it would be sure to be all right.

Beginning to churn again I stared anxiously at the cream. Bill, passing by and hearing my tale burst out laughing, for the cream was now whipping up into a spectacular lather.

"That's all the soap residue in the sump," said Bill. "I'd throw it away if I were you."

"But I can't think how it got out of the churn," I said, "and how it didn't even mark the cushions."

"What you didn't think of," said Bill, "was the hole where the paddle-shaft goes through the lid. And spinning like that I suppose the cream just whizzed past everything."

Only one other time have I seen a half gallon of cream go to waste. I was carrying it up the cellar steps for butter-making, in two basins, one balanced on the top of the other with a plate between. On the top step I put down the basins to open the door, and at once a crisis set in with the swift arrival of all the cats, determined, as usual, to go down into the cellar. With the cats and the cream and the door and the ancient worn steps, somehow the top basin caught the bottom of the door and that unbalanced the lower basin and the whole lot went down the steps.

The basins broke (they were nice blue and white ones) and the cream poured down the pink brick steps in a slow, voluptuous, widening flood, like rich light lava. At the bottom it set out across the cellar floor, spreading and sending out fingers where the bricks were worn down. By this time I had rushed for hot water, soda and a mop, but soon the cream was more than half-way across the large cellar. It was a sort of gourmand's nightmare, mopping up seas of cream which spread the further as one mopped. Seven bucketfuls of water I fetched to it, and still the water was cloudy.

One had a strange illusion of peace and plenty, of a land flowing with milk and honey: "and they shall mop their floors with cream."

As for the cats, they soon lost interest and went out hunting; our Siamese are not very fond of cream. The real attraction in the cellar was the meat safe.

CHAPTER FIFTEEN

New Gardening Policy

Relieved of cow-care and some other duties of the farm, one might have thought that I would have turned my attention to the house, so long in need of a spring clean, new curtains and covers and fresh distemper. In fact, I turned to the garden.

Gardening here, for me, had always been a struggle. There was simply not time for it, and often there was not strength for it. My gardening was a long, losing battle against weeds, animals and growing lawns. What would have become of it if I had not heard of Michael Haworth-Booth I cannot say, but, on the recommendation of our gardening doctor, I bought his book *Effective Flowering Shrubs.* This book provided so beautiful an answer to my particular problems that I began at once to follow the ideas it presented.

The new popularity of flowering trees and shrubs may be well known to many gardeners. I knew, myself, that they were recommended for a beautiful and labour-saving garden, but I had not previously read very much about them. I had never thought of shrubs and trees as being particularly showy, but Mr Haworth-Booth says in his foreword: "I am out for flowers — large, bold, beautiful flowers . . ." and in his following pages he succeeded easily in firing this particular reader, at least, to follow his suggestions.

It did not take long to banish my doubts about effectiveness. I had soon forgotten the flowering currants, buddleias, mauve lilacs and the rest of the dowdy horde of which I had always thought when anyone said shrubs: instead, my mind was full of azaleas, hydrangeas, magnolias, camellias and such aristocrats as those. (Why had I never before discovered that camellias, far from being tender glasshouse plants, will flourish outdoors like a hardy rhododendron?)

The labour-saving aspect appealed to me enormously, and it sounded obvious and simple. With no soft-stemmed plants in the garden, one could mulch the trees and shrubs with several inches of dead leaves or bracken, thus keeping down all but the strong perennial weeds, which must be eliminated first by plain old-fashioned labour.

"What, no weeding to do at all?" my family said disbelievingly.

I said, "None at all, if we keep the mulch thick enough. And no watering either, because the leaves conserve the moisture."

"There's always a snag somewhere in these wonderful new ideas," Bill said. "What about the expense of buying all the shrubs?"

"At least," I said, "we shan't have the same expense again each year, as we did with bedding plants and things like that. You only have to buy a tree once, with any ordinary luck."

We argued about the virtues of composting leaves

instead of packing them down on the borders as they were; but the weed-smothering properties are lost when leaves are made into compost. Probably some other properties are lost too, unless the heap is kept covered from the rain.

I began my gardening all over again, from the beginning. The first thing I bought was a barium soil-testing outfit (quite cheaply, from Boots) to test for acidity, which most trees and shrubs like and some must have: because here in this old garden one is always digging up ancient foundations with lime-mortar rubble all about them. Then came the careful preparing of the ground, the rounding of all avoidable angles and the turfing over of any incidental flowerbeds, so that the restful outline should not be spoilt. My roses had been in such a central plot, but now I dug them all out for replanting among azaleas and hydrangeas, being impressed by Mr Haworth-Booth's condemnation of the "concentration camp existence" in a rose-bed.

The ordering of the vanguard of the garden's new inhabitants was a pleasure only darkened by the need for immediate economy. Beseeching my family never again to give me anything but plant tokens for birthdays and Christmas, I cut down my list to a few indispensables, including some fine hybrid rhododendrons and azaleas and two oriental cherries.

I fell a slave to the exquisite little evergreen Japanese azaleas, which are beautiful all the year round. Some varieties (particularly the soft pink Hinomayo) will flower for five weeks on end in the spring and early summer, and look attractive all the winter with their many-tiered forms and fine foliage. They are among the least expensive of the best flowering shrubs, costing less than a pair of stockings.

Magnolias — like towers of white water lilies — had always appealed to me, and I began with the variety Soulangeana, first testing the soil where I wished to plant it, for these trees cannot bear lime. The oriental cherries are more accommodating and can be planted in very alkaline soils. I chose Tai Haku, the Great White Cherry, perhaps the loveliest of all cherries, though some would put the single pink Sargentii first. Flowering at about the same time as Tai Haku are the dwarf mountain rhododendron Blue Diamond and the soft yellow broom Cytisus Kewensis, both of which I planted close by for an effective three-colour scheme.

When the first planting was completed, Bill and Shelley and I took the tractor and trailer down the lanes and brought back loads of autumn leaves, which I laid down and packed tightly round my plants and over all the dug soil. Nowadays, for lack of time, we make an arrangement with the roadman to deliver loads by lorry, and this has worked very well.

Wind and blackbirds can be a nuisance in disturbing and scattering the leaves. I find that really tight packing helps, especially against the wind, but a few fronds of bracken laid over the top have been very effective. Unfortunately, bracken-cutting and leaf-fall do not coincide; the one must be held until the other is ready.

I began to discover, as many gardeners have discovered before me, the sheer excitement of reading catalogues in bed. No thriller can beat them, for downright evocation and suspense. ("I wonder how it would look near the front gate? . . . really, I oughtn't to afford it . . . but remember the one we saw at Wisley last June! . . . civilized people could consider it an essential to life . . . perhaps I could afford two . . . or even three . . . but what about these? they were glorious at Chelsea . . .") I would be executing ticks and crosses and footnotes in a wavering hand, against the pillow, my nib crossing among the sheets and blankets and Vashti's hairs sticking in the prongs of it.

Chelsea Flower Show! — the undoing of so many conscientious economists, but what fun it is to be undone. Here I was undone in favour of the lovely Chinese-yellow Knap Hill azalea Harvest Moon, which I shall plant against my rhododendron Purple Splendour, and here I fell for another Knap Hill azalea, Satan, a gorgeous wicked red.

One of my gardening mistakes is polygonum bald-schuanicum. Anyone who plants this Jack-and-the-beanstalk climber is almost sure to regret it. Cheered by the catalogue description of this "fast-growing, hardy, floriferous climber" I put it against the bare bleak dry corner of our north-east wall, where presumably all else would die. That was five years ago. Some people call this plant Mind Your Own Business. Now I know why. Within a year it was looking in at everyone's window: now it is up on (and in) the roof tiles, waving along the telephone wires and tickling heads of passers-by. Once I went away for a fortnight and returned to find it curling round my bedroom curtains.

Every autumn I take a long ladder and my secateurs and shears and knife and cut it hard back. But every spring it is blocking up window hinges again, and every summer it has its exquisite hour (in justice to it, the "hour" is weeks and weeks) when it is clouded with waving, trailing, dancing white sprays like an enchanted waterfall. But don't be seduced by its summer glory: don't plant polygonum baldschuanicum — unless you have a great old Nissen hut, or boiler factory, or prison wall in your garden that you must cover up quickly, at any price. And even then, be prepared to let the thing dominate your whole garden, your view, your house (inside and out), and, of course, your own business. With this plant it is all or nothing. I keep trying to make up my mind to cut mine down, but once a thing is planted it takes great strength of mind to destroy it. So don't plant polygonum baldschuanicum.

In its new form, my garden is less than a year old. Many of my plants are flowering now for the first time and all are small and unimpressive. No one coming into the garden and looking round sees it as I see it, in my imagination, and as I hope it will be one day — the trees mature and spreading, the bushes broad and thick, and all the jungly places round the old wing site cleared and cleaned and planted.

This year I have begun on the old wing, getting about half the area roughly levelled and sown to grass, but leaving the inevitable old foundations where they were because the family liked them.

The path, about half finished now, was a new idea. It was one of Mr Haworth-Booth's notions and seems to me so remarkably good and easy that I ought to pass it on. I had for a long time been collecting large flat stones to make my path, which runs between the new lawn on the old wing site and the Japanese azalea bank, but I had been postponing the making because concrete (in which all paths should be laid unless you are willing to spend whole weeks weeding with an old dinner knife) is so tiresome and heavy to mix, cart and handle. I know, I have tried it. It is all that I can do to shift a bucketful, let alone a wheelbarrowful, and if it isn't used quickly it starts to set. I was always being called away in the middle of the work for some urgent business, only to find that the concrete had "gone off" when I returned.

"Why not lay the stones in dry cement and sand?" said Mr Haworth-Booth, arriving one day with a Persian lilac.

I said, "Do you really mean *dry*?"

He did, he said, quite dry. You mixed it up, one part of cement to three of sand, and then put it down three inches thick on your prepared path-site and laid the stones upon it, filling up the cracks, making sure of the level, and then carefully going over the surface with a soft brush. You then watered the whole with the fine rose on the can (I use a spray), put up a little notice to keep folks off for two or three weeks and carry on when you next have time to spare for it.

A whole barrowload of dry cement and sand is light enough for me to wheel about. And if someone suddenly calls, or it starts to rain, I can wheel what is left into the dry and leave it until another day.

I have always felt sorry that I am the only gardening enthusiast in this family. Such a person must always be the butt of everyone else's jokes — like the ardent golfer or the woman who lives for bridge. But I think that gardening, of all the human activities, brings out the best there is in us — though that may not amount to much. Perhaps it is only that when we are doing the things we like to do, we feel (and therefore look) our best.

However it may be, my theory was recently upheld by a neat double-edged compliment from Shelley: "You know, Mother, when you were sharpening those shears in the workshop you looked so nice I thought it was a stranger."

CHAPTER SIXTEEN

Harvest, Haying & a Treasure Hunt

Haytime and harvest at Punch Bowl Farm are arduous occasions for often only family help is available. Everyone who can drive the tractor, use a hayfork or lift a sheaf is in the fields. Last August the oat harvest was a wet one coinciding with Shelley's Bavarian holiday and coming just before the advent of our allies the Webbs, whom we had not seen for two years owing to their having acquired a trailer caravan and taken it first to France and then to Wales. There were only Bill and Shaun and I to stook it all and get it in and stack it; and I wasn't very much use because of my recent illness. It would not have been so bad if it hadn't rained, or at least if it hadn't rained so hard and often. The rain began as soon as the cutting was finished and before we had managed to stook half of the sheaves lying out in the wet stubble. We stooked in the rain, our wrists getting scratched and red with wet straw; the sheaves were very heavy, weighed down with the rain, their soaked heads falling sideways, forwards, all ways, instead of standing stiffly erect to help the stookers.

That harvest was difficult to dry; no town wet washday

could give you any idea of how difficult it is to dry eleven acres of wet corn. Especially when the corn is wet *inside* the stooks, through being heavily rained on just before the stooking. If corn is first stooked dry it will stand a lot of bad weather, shedding the rain from its roof-shaped stooks as a roof will. When you have plenty of labour it is possible to stook as fast as the binder cuts, but last year, with Bill on the binder, there were only Shaun and I to stook.

At last, despairing of ever drying the wettest sheaves, we waited till the weather cleared and settled to fair and then we pulled the stooks apart, loaded them on to the trailer and began distributing them upon the fences, walls, hedges, sheds and low branches round the farm. Like a great silvery-gold wash strung out upon an endless clothes-line the oat sheaves hung there. The harvest lapped the farm all round with tides of yellow like a sorcerer's sea. The long fence running up between the Little Orchard and Barn field was a great pale golden snake, and the air was filled with the damp-sweet smells of drying corn.

When at last the sheaves were dry enough to carry we worked all day until dusk, Shaun driving the tractor, Bill pitching and I standing on the load and building the sheaves. Shaun was then only nine but he worked the same long hours that we did, handling the old tractor expertly in the difficult hilly oatfields (there is not one place on our seventy acres flat enough to make a tennis court) and manoeuvring through gateways and up to the stack like an old hand at the job. It was because of him that we got the harvest in and stacked in fair saleable condition — a thing I had not dared to hope for when I had seen it sodden in the fields, with green mould flying from the touch and long white shoots trailing out inside the stooks.

We cleared the two fields, Barn and Hanger; the stack near the Swing Tree growing higher to its eaves; then we cleared the hung-out sheaves, snatching them from walls and fences where Evil Eye the gander had been gorging himself and his wives on them; and then we took them from the bushes and trees, plucking them from the branches like some great exotic golden fruit.

The cats loved the harvest, when the worst of the wet had dried from it. Coming with us to the fields they killed dangling oat-heads, hunted harvest mice in the tunnels of the stooks, tore after the tractor and were lifted up to ride home on the top of the load.

I missed my mother in those days. An empty house is like a room without a fire, there is no life in it; no one preparing the cool lunch you are longing for, no rattling of teacups to greet you as you plod down the track, no clean dusted rooms with flowers in them. There was no time for anything but the harvest; and it was impossible to get daily help from the village. We had no accommodation for a live-in housekeeper, and in any case we like the house to ourselves. But if you had asked us if we would exchange with anyone in town we would have stared in shocked surprise at such a notion.

We had a visit from strangers to-day (all writers have visits from strangers) and one of them said, "I would give anything to live on a farm like this and be able to work here."

I thought, probably you would; but the romantic town idea of life on farms is far from true. If you really love farming and country life you must love it with all its wet harvests, strayed stock, and bitter losses. (That very morning we had lost twenty-seven chicks out of thirty by rats which had found their way somehow into the chick house.) It is necessary to have the philosophy of the old sheep farmer who said, "It's not so bad; six years out of seven my sheep keep me; one year out of seven I keep my sheep." It was May when he spoke, and because the grass was late he was keeping some six hundred ewes and lambs all out of his own pocket. But he said, "It's not so bad": and of course for real country people it isn't — at least it's not as bad as life in towns!

Towards the end of August the Webbs were with us again, the two boys staying in the farmhouse and Eva and Kenneth in their little trailer caravan. It seemed as if the old farm came to life again; we had been such a small household, with Shelley away in Bavaria and no visitors yet. Far away seemed the days of the first Webb camp, when all of us were winding up the water still, and the house was dark and inconvenient; and Dion and Julian were first learning the jobs of the farm.

Now the boys were toughs of sixteen and eighteen, who thought nothing of a day's tractor driving, bush whacking or tree felling. Bill would tell them which trees were to come down and leave them to it. This year the trees were well-grown elms in the Little Orchard hedge, and though I sought for their reprieve they were doomed because they overshadowed the young fruit trees. The boys decided exactly where the first tree was to fall, and then cut a deep notch with the felling-axe, low down on the trunk at the side where the fall would be. They worked with the cross-cut saw until the trunk was balanced on a sliver, and then shoved against it with their combined weights.

"Timber!" they yelled in the professional manner, though there was no one in the way, and over came the elm tree, cracking like a rifle-shot as the last trunk-wood snapped. Slowly, almost deliberately, the tree came down, crashing and sinking on its branches in the grass. Now the tree had to be cleared of branches with handsaw and billhook, ready for Bill to cut it up on the saw-bench. Some of our wood is cut up into logs for burning in the boiler and inglenook, but never elm-wood. Elm, though a very bad burning wood, is good for building. Our tractor-trailer was made by Bill from a pair of old car-wheels and elm planks felled and sawn on the farm. These elms of Dion's and Julian's were later used for making gates and poultry arks.

After the felling there was a sudden renewed attack of general interest in the old wing site. Talk of the "treasure" and speculations about the fallen wing started it off. The whole household converged on the place with picks and shovels, spades and forks and mattocks. At times there were nine or ten of us there, including Shaun's young friends, David and Anne Brown (then living in a flat in Rock Cottage) and a fourteen-year-old, Mark, who bicycled fifteen miles to join in the excitement. The old hummocky nettle patch, where the wing had once stood, looked like a gold rush. It was almost the last part of the garden still untouched, its heaped-up piles of nettle-smothered rubble having put off all attempts till now. There was also a scaffolding against the north gable on which Eva stood for long periods, decorating with scooped Tudor roses a bar of wet plaster Bill had put in to fill a gap above a beam.

With a little organization, I thought to myself, all this energetic treasure hunting could combine with garden making. Instead of just throwing the rubble about from one place to another, it could be cleared right off the wing. A talk with Bill and Kenneth about this produced, almost at once, the strange sight of the tractor and trailer in the garden. Hacking out the old derelict hedge between the wing site and the orchard, Bill backed the whole outfit down to the excavations, bumping up and down over humps and hollows in his path.

"I'll stick an elm gate up there, when I get a moment to spare," said Bill. "Just now, I'll put this wire across, for a temporary measure."

Rubble flew up on to the trailer, bits of pottery and tile were earnestly examined, and the old wing site began to look like a badly hit bomb-site. A pile of eagerly salvaged oddments of domestic ware slowly accumulated by the house wall; nothing very notable, but enough old glass and tiles and bricks showing marks of burning to convince us more than ever about the fate of the fallen wing. Bill found a fragment of Victorian pottery with a knock-kneed bathing damsel on it in sepia and white. Sticking it up against the north wall, he said it was his pin-up girl and showed it to everyone who called.

I spent my time dithering between being thankful for

the usefulness of the trailer, and furious because of the hens that now swarmed into the garden through the "temporary measure" wire. I can think of a great many things that are more use in a garden than a party of hens, and of few that are more infuriating to a gardener. But as Bill told me, you can't have it all ways.

Then suddenly he announced that his second-cut hay in Little Orchard was fit for carrying, and the old wing excavations were called off for a day. As it happened, that was the day when Vera and Norman Pritchard called to see us. We had known them for a great many years without ever seeing them at the farm, for they also were continental-caravan-holiday enthusiasts. This first time, they called early in the evening on their way back to London from the Farnborough Air Show; and so, as goes without saying, they were hardly dressed for Punch Bowl Farm.

Eva and I decided to take Vera for a walk round the fields, leaving Norman to watch the haying. I felt the haymakers could do without my help, for once, since it was so small a field and the labourers were so many. Coming back along the yew tree track, with Vera's nylons no longer what they were, we could already hear the sounds of cheerful activity in the hayfield just across the little valley. The yew tree passed, we had the whole scene laid before us. The farmhouse stands, of course, in its own small valley, and on the eastern slope of the valley is Little Orchard. The yew tree crowns the western slope (which is steeper) and it was there that we stood, looking across the garden and past the roof-top to the haying in Little Orchard.

There was a low westering sun that missed the valley now but lit the hayfield like a floodlight at a pageant. Shaun was driving the tractor among the little staked trees (anxiety for him suddenly clutched me — the slope was so steep) and the top of the field was full of people cocking, raking and pitching hay. Shirts had been long abandoned and brown torsos moved beneath loaded hayforks — but who was that stranger? standing half-buried on the top of the trailer-load and building the pitched hay.

"Now just look at Norman!" said Vera. "Isn't that exactly like him? And his town clothes hung all over the fence!"

The haymakers heard our voices and waved. We waved back. Their voices and our voices carried clearly on the evening air, and the scent of hay was everywhere. I looked and listened to it all for a few minutes, making the most of a day when I could simply stand and watch instead of working. It was the only haying that I have ever observed from a distance since we came to live at the farm.

Vera and Norman went back to town that night; but a few days later they returned with three of their children: Robert, sixteen; Gail, fourteen; and Dion, twelve (the eldest, Elizabeth, being too preoccupied with her approaching marriage to leave town). As you may have guessed, this Dion is a namesake of the older Dion; they are distinguished by the prefixes "old" and "young."

A great deal was crammed into that day. Bill rediscovered a set of small-gauge rails and trucks which he had stored away, and laid them out along the length of the old wing. There was no engine — it was the only thing lacking — but the little railway was a perfect system for clearing the old wing; it was so much easier to handle than the tractor and trailer. It's rails spanned the hollows and side-tracked the humps, the little trucks rolled down them full of rubble going out, and full of people coming back. The railway was quite the most popular thing on the farm at that time, and it helped more than anything else to co-ordinate treasure hunting with garden making. Admittedly the working speed grew faster and faster, until people were being spilled out of trucks in all directions, but this only made the fun more furious; and it didn't affect the precision with which rubble was dumped in one place, broken bricks in another, and whole bricks neatly stacked in another.

By the time the gang was called off for a large combined picnic lunch the wing was almost levelled. Fowls had been going in and out all the time, scratching at the roots of all my treasures, but this awfulness was part of the price of clearing the old wing for the lawn I meant to sow next spring.

As soon as lunch was out of the way Bill fetched the tractor and trailer.

"Anybody want a ride round the farm?"

At once nine people crammed themselves on to the little trailer, which was without its sides to make the trip more exciting. Eva, Vera and I waved them off, and then set out to walk across the fields to see what could be seen of them in passing. Being path-bound because of stockings and bare legs, we did not see a great deal of them, the tractor taking a cross-country route that would have blenched a steeplechase.

"It's all rough ground," said Bill cheerfully, cutting across our path in Upper Six Acres. "All got to be bulldozed some day"; and across the bracken and brambles they roared, forging through gorse thickets and sweeping under low branches. Most of the crew were standing now, with the idea of seeing who could remain upright the longest. Now and again four or five would suddenly subside, as if through holes in the bottom of the trailer; and once or twice (when Shaun was driving) the whole outfit was taken under branches that swept all standing survivors straight off into the bracken.

A later glimpse revealed Gail driving. Someone must have told her that the tractor would go through or over anything, because she drove head-on into an oak tree. Everything came to a stop very suddenly that time and the entire company was prostrated. In Nameless field the bracken was so tall that they were completely submerged, only the tractor's air-pipe sticking up above the surface like a periscope, marking their erratic voyage through the greenery.

When we three matrons had finished our walk, and had been talking at home for a long time afterwards about ideas Bill and I were brewing for clearing Upper Six Acres in the spring, the trailer riders were still roaring crazily over the fields. The distant throb of the tractor came down on the wind to us, where we sat inside the Webb's caravan.

That evening I wrote a long letter to Shelley in Bavaria:

". . . Nothing really exciting has been dug up on the old wing yet, but everyone seems to be having a wonderful time. I wish you were here."

Perhaps I didn't, in fact: she was having such a wonderful time herself — seven weeks in an old Bavarian farmhouse, where the eaves are carved, and the animals live under the same roof, and the men wear leather shorts, and the mountains are all around one. And she was, bless her, paying her own way by teaching English to the family: we had only her fare to manage.

"We will have the Pritchards to stay here, perhaps, when you come home," I wrote.

We did, of course. They met her in London on her return, kept her for a night and brought her down to the farm. But before all that could happen, excitement broke out on the old wing.

The very day before the Webbs were to leave us, they struck the original foundations.

"I think there's something large under here," Julian said, shovelling in a spot by himself. "I'm sure there is — large and solid."

Everyone went over to help him, and slowly there came to light a massive stone. Excitement rose to fever pitch, earth flying (to this day, on that spot, topsoil and subsoil are so mixed that the lawn is thin there), and one after the other the line of heavy foundation stones was revealed. Each was tenderly brushed clear of earth and mortar by Eva and me as it was left behind by the diggers, who were soon shouting their excitement over a new discovery — a cobbled frontage.

Daylight failed as the time wore on, and Bill hung an electric light out from the boiler-room window, stringing it up on a hastily improvised swinging framework. Eva and I stood beneath it, huddled against the wall because of the autumn chill and a settling drizzle, but the excavators didn't care about the weather.

"The treasure any minute, now!" they said, and dug on, wet and weary. "We must find it before we go, and we're leaving after breakfast."

The treasure, alas, was never found — as we hardly expected it would be — but a last and crowning discovery was the stump of a buried post, that we took to be a king-post for the vanished room. It was exactly central, and it bore heavy marks of burning.

I wrote to Shelley: "Such a lot to show you when you come back."

That summer the farmhouse seemed more crowded than ever before. With the three young Pritchards staying here and all the Webbs (the boys inside the farmhouse), as well as drifts of casual visitors, we were sometimes as many as eighteen people in and around the house. The summer went slowly out with a blaze of glorious warm weather; the Webbs postponed their departure and every day the garden was the scene of large delightful lawn teas. At that time Coronet was our only milking cow, and she was the provider of bowls of whipped cream and plates of rich butter that made the plainest teas seem festive.

"Next year," we said, "we may have more cows, and be selling milk commercially, and then there won't be all this cream and butter; so make the most of it while you can."

Washing-up afterwards, I am pleased to relate, was strictly shared out so that, paradoxically, the greater our numbers the less I had to do.

Everyone who could wield an axe, a bush-knife or a saw was in the gang that cleared the copse in Lower Six Acres that September. This copse was really an overgrown, spread hedge (called a "rew" in these parts) and most of the trees in it were saplings, gradually invading the arable. Where Bill found so many suitable implements I cannot imagine, but the long copse was so thick with lumberjacks and jills — there were extra visitors that day — that it became almost dangerous to linger there. At least, it looked dangerous, but Bill had his gang very well schooled. No one brought down a tree without fair warning and no one was hurt.

"Timber!" they shouted, and the ground was cleared in an instant.

When the taller trees were all down and only the spindly saplings left, a lot of foolery went on.

"Timber!" Old Dion would bawl, felling a clothes'-prop-sized pole with much flourish.

"Timber!" yelled Young Dion, bending a sappy hazel over to the ground, and, "Timber!" cried Julian tottering past with a little pole, like a small caber, balanced on his hands.

Gail with a bush-knife was someone to give a wide berth; but Robert was the surprising one. Slim, very tall and rather quiet, he is the kind that suddenly astonishes with twice as much strength and staying power as one would have thought possible.

Most of the cats were up there, disapproving of the

ruin in their best rabbiting ground; but when Vera and Eva and I went down to start organizing the feeding of the multitude the cats all came too. They were always well lined up at tea-time because, though not really enthusiastic about cream, they never miss a chance to lick the cream whisk.

That Michaelmas Shaun went away to his preparatory school. He spent most of his first term in bed with mumps and most of his second in bed with chicken-pox, or so it seemed. Visiting was naturally cancelled during the whole of the quarantine period so we didn't see a lot of him. During mumps we had a letter saying "mumps is terrific" but during chicken-pox we didn't have a letter at all, probably because none of us at the farm had ever had the disease. It is easy to imagine, therefore, our joy on seeing again a letter with the familiar postmark and in his handwriting. Tearing it open, longing for news, I found a whole page written half in morse and half in semaphore. Except for "Dear Mother" there was not a word in straightforward English.

Here was I, starved for news of my youngest dear child and faced with half an hour's serious application with code-keys before I could make clear the gist. And were there any code-keys in the house? How I wished I had kept up these youthful arts, as I laid the letter aside and got on with the morning's urgencies until such time as I could settle to the letter.

It was as well that I remembered suddenly the semaphore key in Arthur Ransome's *Winter Holiday*. Now at least I could decode half the letter. Then I remembered seeing a morse key, pencilled and very dim, among some books in Shaun's room. Finding both keys I settled down at my desk with pencil and pad, laboriously picking out the letters. In the morse key the O and K were so dim and undecipherable that I had to put whenever encountering them, "O or try K" and see how the sense of things worked out.

He was all right, he said. He was out of the san. Chicken-pox was all right. He hoped I was all right. Smith, J. B., was still his best friend and would we please send his boomerang.

CHAPTER SEVENTEEN

Professional Gent

Siamese cats have queer effects on people, but usually they are strong effects. One doesn't often see a lukewarm reaction.

"Oh, how perfectly hideous! just like monkeys."

"What exquisite creatures! they're so svelte and dignified."

"Give me a nice fluffy Persian any day; these remind me of rats."

"The thing about Siamese is that you can appreciate their lovely line; it isn't all muffled up with long hair."

"What a terrible squint!"

"Oh, what glorious blue eyes!"

And so on.

Or even, as a lorry driver said one day, having met first Vashti at the gate, then Thistledown in the garden and finally Freyni at the door which I had opened to his knock: "Gor blimey! there's another of them funny cats. They squint at you, don't they."

Or again, as the Nasty Customer said on edging his unwelcome body into the kitchen when I was alone in the place: "Aw, look at them all sitting on the stove-top! Is it true they're savage?"

This was a gift into my lap, so considering the circumstances I felt justified in replying, "Well, they won't go for you if you don't stare at them or move suddenly in any way. But they don't really like strangers very much."

He didn't wait.

Whatever one thinks about Siamese cats one has to admit that they are arresting. Few creatures have that rare and lovely colour mutation of dark points with a light body tone. Jersey cattle have it and they are arresting too; nearly everyone looks twice at a Jersey. Giant Pandas have it in their own form, and Ming was the most popular animal in the London Zoo for all the time that she was there.

Besides being unlike ordinary cats to look at, Siamese have various characteristics unique to their race. They drink a great deal of water, and often like to get right into water. They like lying on high places with their legs hanging down; and all kinds of movement, swinging, riding and sliding. They love long walks; they will retrieve readily; and they almost always reply to remarks addressed to them. Their voices are quite unlike those of any other cat. They are capable of a very great devotion, usually to one person, and they have queer fancies about food. (Cheetah enjoys raw cabbage and simply dotes on pig-meal, wet or dry; both Simbas had a passion for cake, Vashti likes rubber; and they all adore bacon rind, though not bacon.)

Freyni is even allergic, a most human characteristic. (Do cats think, I wonder, when pleased with humans, "Really, they are quite feline!" or cows, "My dear, sometimes they are almost bovine.") We are not exactly sure what Freyni is allergic to, but it is easy to see she is allergic, for she has bouts of coughing and sneezing at times that can be heard all over the house. She is, I haste to add, in the best of health, and the allergy has gone on since her kittenhood. Probably, Shelley says, she is allergic to cats, as my mother is, which must be a terrible fate, because to be allergic to oneself is an inescapable sort of doom. She is certainly allergic to purring, because loud prolonged cheerfulness of this kind always sets her off: and this is a very sad fate too, because she is a cheerful cat and inclined to song.

When Freyni has her sneezing bouts she is very funny to watch; though of course we don't let amusement dull our sympathy for her. But she is so comically rough with her nose. She hits it, shoves it, swipes it aside, blowing down it hard as she does so.

Thistledown is the eternal Delilah. She is exquisite and knows it. Where she is blonde she is unusually blonde, and where she is dark she is black: her contrasts are dramatic and her eyes are slanting and deep blue. She is feather light (hence her name) with small precise paws, and with movements more fluid than a cat's movements usually are, which is saying a great deal. Apart from a few Arab horses she is the only animal I know who obviously shows off. She walks through a room with such "oomph" in every step that one is forced to look at her, if only to laugh, and she sits about in advantageous places blinking slowly.

Thistledown takes motherhood much as you would expect in one of her type. She puts her feet up regularly early on — but yes, I mean this literally. Lying luxuriously close against the Aga on her back, she puts her four small feet up high on the warm front of the cooker. Then after this daily rest comes the fuss about booking her bed. For weeks she has been anxiously wandering about saying "Any moment now!" and "Where on earth am I to have them?" We have a very good cats' maternity ward in the shape of a large washing copper which we took out of its place when we came here with the idea of polishing it and using it for logs. For years now it has been a maternity ward and it makes a very suitable one as it is nicely dim inside and much too deep for the kittens to climb out. In the end Thistle does have her family there but she never goes in until the very last moment. The one place where she wants to have her kittens is in the various pigeon-holes of my desk. If ever (and it is often) I forget to shut the roll-top at these times I can expect to find all my letters and odd papers pulled out and thrown about the desk.

Thistledown, like most Siamese, is uncommonly skilled at opening latches, catches and handles. Doors are nothing to her. She can get into any cupboard and the kitchen door was so easy she would do it just for fun. I say "was" because we were recently driven into fitting a new latch. Being woken at 2 a.m. by Thistle calling the other cats up the stairs behind her was not funny after the first dozen times. She may be as light as her namesake but she can sound like a heifer on the stairs in the deep hours of the night.

Her system was to climb up the door-post and pat the thumb-latch down, pulling open the door at the same time. We used to slip a little wedge under the latch last thing at night and this would stop her, though there was always the penetrating sound of the latch being rattled under that furious little paw. The real trouble about this idea was, though, that people would forget to slip the wedge in place. I tied it to the latch so that it clanked when the door was opened, but still it was often overlooked. Thistledown never lost a chance; if that door was not wedged she would go through it.

One morning after such a night I went despairingly to the phone and asked our local ironmonger's advice. What about an ordinary knob? I asked.

They were sometimes difficult, he said, to fit on to a very old door. He thought really a ring-latch would be the thing, 'm.

I said I didn't know what a ring-latch was, but if he thought it was Siamese cat proof, and easy to screw on (for I should very likely have to do it myself), I would have it.

Very well, 'm, he said, he would send it in the morning.

It was a handsome latch, made of sturdy iron which looked wrought but must I suppose have been cast, because it was so cheap. I fetched my own personal screwdriver from the dresser cupboard and began. The old latch was easy to remove but when I started fitting the new one I ran into such a fog of trouble (nothing meeting anywhere and a great hole to make in the door, as well as another great hole — in the wrong place — to conceal) that Robbie gently took it all from my hands and got on with it. When I saw what he was doing with it later (shortening the iron shaft, cutting and finishing a piece of wood to "bring the door-post out," drilling holes for screws and the like), I was glad I had not insisted on being independent.

The finished job looked very nice, the ring (like an iron slave-bangle) hanging down on the door with a grace that no thumb-latch can equal.

In two nights Thistle had the hang of it. Oh well, it was easy really, she more or less said modestly, all you do is, you climb up the door-post as before but instead of banging on the thumb-latch you just curl your paw into the ring and draw it towards you.

It was then that I had a visitation from my guardian angel, who put the simple idea into my head; which was, to turn the ring *upwards*, so that pulling it sideways could only shut the door tighter.

It works quite beautifully. But, Bill says depressingly, that is only because it is new and so too stiff for a cat to pull the ring down again.

Both the Simbas were expert door-openers, and Dromore Cheetah has already started on the nursery slopes as I shall shortly relate.

Cheetah, of course, should never have come. I have explained how we decided never to have another entire male, and both Thistledown and Freyni were taken to "proper" studs last winter. This is an awful business: shutting the queen in the bathroom for three loud days, because if you take them too early they "go off"; then the car journey to some place half across the county; then three days later the same journey to fetch them home, and if they haven't "taken" the whole thing all over again, which happened to us twice running. With a resident stud all is so much easier. You just do nothing. He sees to it all. But this is the only favourable thing about keeping your own male. The things against doing so are; spraying in the house (often a drawback, but Cheetah never does this), fighting (male Siamese are often very ferocious and may decimate a whole neighbourhood of toms, which doesn't make you very popular), wandering, with its dangers of traps, guns and cat thieves, and the fact that he may be away on business somewhere at the very time you want him.

Cheetah, a blue-point, was offered to us as a gift by his breeder, Mrs Williamson, the same who bred Vashti and Freyni. He was nearly a year old and had been made much of, all his life. He had, said Mrs Williamson, great charm. He was, in fact, a gent. His destiny would ordinarily be the same as that of nearly all Siamese stud cats, a lazy, safe, luxurious life in a cage. Admittedly a Siamese stud's cage is usually a grand heated house with large wired-in private garden often complete with a useful tree, but it is still a cage. Cheetah, said my friend, had never been used to confinement. The more she thought about it the more she thought that freedom was worth its price in dangers to a Siamese gent, as much as ever it was to a human. Would we like to have him as a gift?

"Well, you know about both our Simbas," I said doubtfully. She did, of course, but there it was — better perhaps to meet a sudden end one day after a life of perfect freedom, than to die of bored senility behind wire-netting.

At the time of writing we have had Dromore Cheetah for a little over a month. He hated the journey, never speaking a word, and then for a couple of days he lived in the bathroom, worrying us acutely because he wouldn't eat a thing. No, he said, no, he just didn't feel like it. We tried him with cooked meat, raw meat, dried meat, tinned meat, sardines, cream, and milky porridge, all in vain. He graduated from the bathroom to the downstairs rooms, where Freyni was so ferociously hostile to him that we had to keep them apart. Vashti declined to see him at all; he "wasn't there." But Thistledown, true to character, just loved him from first sight. She was at that time in the "any moment now" stage, but setting aside personal anxieties she used all her manifold charms to ease the homesickness of Cheetah. Seeing them curled up together, washing behind each other's ears, you would have thought they had known each other for years.

Feeling worried about Cheetah's refusal to eat I took the little gun and set off up the fields to try to get him a rabbit. At any other time, I suppose, I might have got one, but not now. At the first useful rabbit corner I encountered Bill, contemplating a rough field destined for clearing. At the next useful place, what should be there but a large fox, lying indolently in the middle of the path. At the next I surprised a hunting cat exploiting the corner before me. Not a rabbit had I seen up to now; but in Lower Naps, on my way back home, there was a beauty. It was sitting there with its back to me, the wind was right and the range was right. But it was sitting by Nanti's feet. Not even for Cheetah was I going to risk shooting my mare. In a moment she saw me and whinnied, and the rabbit rushed away, carefully keeping Nanti between us as it did so.

I saw nothing else of note, except a mole on the ploughland who, on hearing me, just quietly submerged. There was no fuss, no flapping, no rotary-digger business. The earth might have been water and the mole a floating fish. It was a charming sight but not much help to Cheetah.

In the end, since he grew playful before he grew hungry, we got him started by throwing bits of meat for him to catch and personally kill. We found that this way he ate every one, and once started the rest was easy.

As time passed, and largely due to Thistle's sweetness, Cheetah settled down. Vashti gradually permitted herself to notice him, in a distant chilly way, and even Freyni merely insulted him instead of rushing to plant a black eye. The full charm of his character now began to unfold. To begin with, his looks: he is just like a polar bear cub, large and heavy and rather clumsy and very nice to pick up and cuddle. He purrs like a waiting tractor, in loud syncopated bursts, interphrased with lion-like snorts and snuffles. He will cast himself down bodily on people's feet, stretching elegantly and turning up his slow blue blink. He is bear-like in action as well as in looks, knocking things down with a sweep of his tail, slipping off things and bumping into things. He talks a fair amount, though usually only when spoken to, and he has more or less told me that he is really a home-loving cat, not at all fitted for the life of a professional gent, but he is following in the steps of his father (who was in the same vocation) and of his father before him. Whenever he talks his grey lion-like muzzle wrinkles most engagingly over his nose; it is as charming a gesture in a cat as crinkly eyes are in a human smile. He has two voices; one, the blood-chilling challenging howl of the entire male Siamese, intended to carry over half a neighbourhood; the other, a soft effeminate comment intended only for family listening. It is comical to hear him, howling his proud arrogance at shouting pitch from the high yew tree bank; then, when Shelley or I call him, to hear his voice at once drop down in tone and in volume, transforming itself in a breath to the gentle croon he keeps for us.

His first experiment in door opening was, unfortunately, with the Bendix. I had left it earnestly washing and spinning and rinsing and all that kind of thing while I went out to do some gardening. Cheetah had been sitting viewing it, as one might television, his head going round and round with the whirling washing.

"Do you know the kitchen's full of water?" Shelley asked me, coming down the path. "Daddy says the Bendix door was open, and he's shut it."

Rushing in I saw the Great Flood, and rushing out I fetched the mop and bucket.

I am sorry to think it, but it seems that Cheetah must be one of those who have to see what makes it go round.

He hasn't thought about hunting very much in the month since he came here from his town house. The fields, woods and buildings have been enough to absorb his interest by themselves. One day I found him vaguely blundering in and out of my sweet williams after a chick of Bill's that had escaped. (I can't think why, but everything on Punch Bowl Farm escapes.) Snatching him away first, I then restored the chick to its run. Cheetah, amiable as ever, said how sorry he was, he didn't realize it was one of ours, or words to that effect. The next day I saw him standing watching a bold blackbird who was pitching my leaf-mulch out on to the lawn. He could have chased that blackbird and been welcome, but he stood there, twitching his blue-grey ears uncertainly, then turned and walked on down the path. Probably one of theirs, you never know, it seems, I think he said.

The other three cats hunt perpetually. You can usually guess what they have caught, even if you aren't in at the death, either by inspecting the horrid internal leftovers or by looking at the hunter's figure afterwards. A really circular cat always means a fair-sized rabbit.

"Wouldn't it be fun if we looked like that after meals," Shelley said one day. "You could say when someone called, 'I see you have already lunched, Mrs So-and-so, or we would ask you to join us.'"

Cheetah gave me a bad shock one day when I was writing at my desk by the sitting-room window. I had heard him trotting heavily about on the stairs and in my room directly overhead, and he had spoken once or twice about the annoyance of having got himself shut up inside the house. I was intending to get up and let him out, but as I was busy the moment kept getting put off. The corner of my eye then suddenly registered the impression of a wonderful outstretched flying cat, sailing past the window at my side. It wasn't possible, I thought anxiously, but I knew it was and hurried outdoors. Bored with waiting, Cheetah had made his exit from my bedroom window, which is sixteen feet from the garden at the least. Expecting to find him felled with a broken leg or worse, I ran round the corner of the house. Cheetah was strolling over the lawn to meet me, saying something to the effect that I needn't have got up now, I was too late to be of any use; and he strolled on past the lilacs and down the steps by the creeping sun roses, and all his legs were as good as ever in his life.

All the Siamese cats dislike recorder music, which is a pity because Shelley plays this instrument quite well. Having taught herself on a cheap plastic model, years ago, she now has a lovely hand-made tropical rosewood recorder on which I like to hear her play. The cats, however, always go out one by one, like the players in Haydn's Farewell Symphony. First Catto will be at the door, ears twitching painfully, and three minutes after we have let her out it will be Freyni. Thistledown, having borne the agony for two minutes longer than this, will now rush to the door; followed finally by Cheetah just when we have settled down again. But he is nothing if not courteous. His tail waves apologetically as he looks up politely at the door. "If you don't mind," I think he says, "I really must be going now; it's been so pleasant seeing you."

CHAPTER EIGHTEEN

Burnings, Pigs & Skins

Through the winter we discussed plans and ideas for clearing Upper Six Acres, one of the rough fields. This field had been chosen, after inspection walks and discussions with the Surrey A.E.C., for three reasons: because it was the flattest of the rough fields, making cultivations easier; because the soil samples taken showed the soil, though poor, to be no poorer than elsewhere; and because it was one of three fields adjoining those already in cultivation. The other two fields in this row, Hunter's field and Upper Naps, are both so hilly that, all other things being fairly equal, the choice fell naturally on Upper Six Acres.

The acidity of all the rough land proved very low indeed –pH 4 to 4.5; perfect for rhododendrons and azaleas, I reflected wistfully, but very bad for farming — which indicated heavy dressings of lime before sowing.

Upper Six Acres was in that state of reversion which is not quite far enough gone to call it a wood and yet too far gone to call it a field. There were trees thirty feet high in it, and bushes of gorse and broom and bramble higher than a man. It was covered with heather and bracken, and picking blackberries in it had always been a hazard with torn stockings, feet disappearing down hidden rabbit holes and once the picker almost vanishing entirely in an unsuspected old bomb crater.

We considered tenders from contractors for bulldozing and clearing Upper Six Acres, but Bill decided to do it with labour employed by himself. Before anyone else touched the field Bill began with the preliminary burning, late in March, after a wide firebreak had been cut and ploughed round all the borders. A windless day was chosen and, taking a blowlamp, Bill fired the field from all sides. The weather was dry and it was too early in the year for much sap to be running; conditions were almost perfect for burning. I went up to the field later in the day, when the fire was well established, and long before I reached it I could see above the hedges the rolls of curdling orange-grey smoke. In the still air the smoke floated high, forming a strange brown cloud that mingled slowly with pure white vapour trails from passing aircraft.

When I reached the gateway I stood for a minute staring at the extraordinary sight of leaning smoke columns, thick and curling though there was no wind at all; and the smoke all leaned towards the middle of the field as if sucked there by a central vacuum, the opposing smoke-streams reminding one of steamers passing on a calm day at sea. Under and in the smoke the red fire travelled with a noise like rushing wind; it leapt with a liquid movement like a pouring upwards, sending out loud cracking and crackling noises, sizzling like milk boiling over when gorse clumps caught alight.

Tall birches in the fire's path stood breathlessly still. The fire reaching them roared under them, and they leapt to life, swaying and tossing above the flames as if in a high wind. This was to me the most amazing sight of the burning.

In the gorse thickets the flames sprang up the bare stems, making flaring crackling torches at the tops. Long stems burned like brands; the burning in the heather curved out like a black tide fringed with foam of fire; then the fire passed on, leaving behind it the great black waste which was one day to be a cornfield.

I could see Bill, watchful with his rake along the boundaries; and everywhere the burned field lay under feathery black and silver ash.

Within a week or two the holidays began, and with them the arrival first of the Pritchards and afterwards the Webbs. Upper Six Acres absorbed everyone through the whole of these Easter holidays. There was still much burning to be done where patches had been circled by the fire, and there were all the dead trees to be cut down and their stumps winched out. Then there was the first deep ploughing. Old clothes were worn and every night the bath would receive black bodies covered with soot and ashes.

The boys tackled everything. Whenever I walked up to the field I would see, perhaps, Young Dion, Shelley, Gail and Shaun absorbed in burning, Julian and Old Dion wielding felling axes, Robbie driving a borrowed tractor with a trailer-load of tree-stumps, and Bill driving the other with a strong chain attached which Kenneth was hooking round tree-stumps ready for the tractor to haul them out.

No one was civilized in the burnt field. Smutty faces, black-streaked trousers, tousled hair — we all looked alike. And there were no union hours; while the light lasted the work went on. Walking up there one evening after dusk, Eva and I could see the dim outlines of Bill and Kenneth with the tractor still hauling out stumps, and the younger gang piling up their great bonfire of branches and scrub. Only Young Dion still worked alone, absorbed with burning. His bent figure, lit up by his blowlamp and the curving bow of flames, looked black against the flushed grey smoke around him.

"Think of it," said Kenneth, pausing to notice us, "a boy being encouraged to burn things. You know, it's really rather fun for them."

It was fun, of course, just as the reverse applies to country people to whom a day in town is a great excitement. With me one day in town at a time is enough and I soon long to see the farm again, but when Vera suddenly says, "Now how about a night up at our place? We could do a theatre and have dinner out"; or "Supposing you come back with us and see Chelsea Flower Show? the farm won't miss you for a day"; then I am tempted to drop all and go.

The plough was in Upper Six Acres before the burners had left it, cleaving its deep double furrow with a smoke-screen thrown across its path. As soon as a section was ploughed it was ridged and planted to potatoes for the pigs, though more than half the field was yet uncleared of stumps and roots.

These pigs, four Wessex Saddleback gilts (which in pigs means a maiden), were a new line for us, arriving just after the great burning. It is a good thing to buy livestock when possible from friendly neighbours, and these four came from Ridgeway Farm, where Shaun had walked each morning in his Godalming schooldays for the six-mile drive with Mr Jupe. Of these pigs one was nearly all black and was unimaginatively called Blackie by Bill. The others were of orthodox saddleback colouring and I called them Dingle, Bellybody and Cuss (after a genuine old-established firm, believe it or not). Certainly any of them could justly have borne the middle name of the three, and all of them were cussed for escaping many times, and Dingle is as good a name as any for a pig.

There is a saying about pigs which goes like this: "A dog looks up to you, a cat looks down on you, but a pig is the only animal that treats you like another human being."

Our pigs became such adepts at escaping that I began to worry again about the garden. I had just had Clover and Coronet trampling all over the newly sown lawn (and after rain) and though the pigs were a long way from the house they can travel incredibly fast and nimbly. So Bill put a branch of the electric fence all round their run. No one saw their first introduction to it as Bill had to rush down to the house to switch it on (it is a mains fence, not a battery one), but when he got back to the pig-run they were all huddled in the middle of the run, facing outwards and staring at the fence. I hope they remain as awed as this. I did once hear about a great fat sow which used to scratch her back on the live wire of an electric fence; every time the electric pulse went through the wire (which it does at second intervals) she gave a little shrug, like a hiccup, and went on scratching.

Bill and Gail and Shelley, with Young Dion and Robbie, fixed up the pig-house and large run in an unburnt corner of Upper Six Acres so that the gilts could help in the clearance. All pigs, I am told, scratch themselves. I am assured it doesn't necessarily mean a rush for the D.D.T. But no one, I am sure, would believe how strenuously they scratch unless they have themselves observed the scratching. One day I was walking down the field with Vashti and we both noticed a terrific thrashing and swaying of the trees in the pig-run. Vashti stared, her tail beginning to fluff out. I stared too, wondering what strange creatures could be leaping around in the branches. It was a long time before, coming nearer, we realized that it was not creatures in the branches at all but merely the four gilts heaving their ample behinds up and down against the trunks.

Any evening in the holidays now one might see Robbie and Bill, their figures silhouetted against the evening sky as they walked along the high yew tree track to the pig-run in Upper Six Acres, Robbie swinging water buckets on a shoulder yoke and Bill with a great sack of lawn mowings over his shoulder. Pigs love lawn mowings and, happily, they do the pigs good. They also do me good, because with a little encouragement and a few comments on how well the gilts are looking on the grass, I can get all my lawns mown without as much as fetching out the mower.

The two Dions and Julian and Robert are all by now accustomed to handling our shotguns, having been carefully instructed and solemnly warned by Bill. During these Easter holidays there was a craze for curing skins. Young Dion shot a large stoat early one morning and it was this that started the fashion. Next there was a mole, upheaving the new lawn on the levelled old wing site. Seeing the earth actually lifting, Bill took a gun and shot into the ground. Old Dion bagged the moleskin and began at once with knife and alum and saltpetre to cure it for making a purse. Then Robert killed an adder up on the high land when they were all working to clear a way for ploughing in Upper Six Acres. First he stunned it and then pinned it down with a forked stick while he cut off its head. He pushed it into his pocket but told me afterwards: "It's wonderful, their reaction: it kept climbing out, so I had to cork it in with a stone."

Later in the day I saw all seven of the young folk leaning over a bench outside the workshop where Robbie was skinning and dissecting his adder.

Even Vashti, to her surprise and fury, was temporarily robbed of a rabbit she had just caught, while someone carefully skinned it before returning it to her.

During the dark spring evenings, after a day of varied farm enterprises, the seven would go to the tottering old barn and there play "Murder" till I called them back for bed. Shaun, at nine, was reckoned old enough now to join in and have a late bedtime, once in a while. Bill and the parents and I, in the farmhouse or buildings (there were enough young folk to play "Murder" without us), would hear from time to time such hackle-raising shrieks from the dark barn that one would think corpses must be littering the hay. And so they probably were.

After the visitors had all returned home at the end of these holidays I had a letter from Vera:

"Dion shot a pigeon in the garden with a friend's airgun," she wrote; "he plucked it, stuffed it and roasted it and had it for supper. Next time it was a grey squirrel, but no one cared to eat that!"

CHAPTER NINETEEN

The Hut

One of the problems of my daily life, especially in the school holidays, is finding a place quiet enough and solitary enough for writing and typing. This problem was much eased by the blessed gift from Mrs Fisher of the writing hut used by her late husband, H. A. L. Fisher, former minister of education and famous author of the well-known *A History of Europe.* The hut, as you can see, is suffering a comedown in its later years, but it is the nicest present I have ever had and is my constant refuge.

All kinds of things can happen when writing in the house. People come and telephone just by my elbow, anybody calls me for almost anything, the cats come in and bang the typewriter keys, someone else has been using my typewriter and left the ribbon turned to red, so that I start off in technicolor, or they have left the shift-key down and I start in capital letters; and, worst of all, I can hear when things are going wrong and naturally feel bound to investigate. When I am up in Lower Six Acres in my hut the whole house could burn down without my knowledge.

I was writing once in the farmhouse because the cows were in Lower Six Acres and I thought they might hang around curiously. In five minutes Shaun was at my side, a boarding-school boy now and, strangely enough, fascinated by Latin.

"Mother, would you like to translate some Latin?"

"I've never done Latin."

"Well honestly, Mother, you ought to know Latin. Look, I'll show you. Here's a simple sentence, 'The natives are preparing an ambush on the troops.' Now *here* I'm writing all the Latin words you'll need, and you only have to put them together; do you see?"

"Oh well, all right. But I'm awfully busy." Long pause with Shaun encouraging me over my shoulder. "Now, how's that?"

"Oh, Mother! Surely you know that the verb always comes last? Can you do it again, remembering that? Then I'll give you another."

I did it again and was flattered to get a large tick. "But this really is the last, old man. I'm too busy. Later, perhaps. Read a book or something, if you must be in the room."

Five minutes later: "I say, Mother, do you know what's an absolute utter chizz about the law?"

"No."

"Well, you've talked to me about it, and it was in the paper. You know."

"The thing is, I'm not really sure what a chizz is."

"Well it's just a *chizz* — you know, a sort of swizzle. The one I thought of was people being whanged into prison when it wasn't their fault, and then when the law finds out it wasn't, and they let them go, they don't give them anything to make up for it."

"Yes," I said, "you're right, it does seem wrong. Now look, old chap, I'd like to stay but I've got to go up to my hut for a bit. Don't let anyone disturb me, will you?"

And so, as ever, the hut is my refuge.

I don't often try to work on books in the holidays, but sometimes there is one that must be finished. It is then that I am most grateful for the hut.

Arriving into it on this day I dumped my books, sun-glasses and pen on one chair, sat in the other and fixed up the ingenious little table arrangement made by Bill, which rests on the arms of my folding chair and is held in place by catches underneath.

Across the open front of the hut was a large branch, put there by Bill to keep the cattle out. I had crawled under this to get inside and now sat looking at the dead twigs, trying to pick up threads in my last chapter. It is a queer thing, but when a lovely view is interrupted by something uncomely in the foreground, it is on the uncomely thing that the eye keeps coming to rest.

I began to write, and at once heard a breathing in the doorway. Looking up I saw Coronet. She was staring at me through the bare branch, delighted at the prospect of human company.

I said, "Hullo, Coronet," and went on writing. She drew nearer and soon was head and shoulders through the branch, her long sentimental face hanging just beyond my pen, dribbling and slavering on to the wooden floor of the hut. Reaching further she began to rub her face against my table, her eyes shut, long black lashes lying on her cheeks like a baby's. I gave her face a firm shove, but it was no good; I had to undo my table and let myself out to go and rearrange the branch.

For the next few minutes Coronet stood there, ogling me with large dark eyes, snorting comfortably to herself and belching and rumbling happily with all her four stomachs. Her wet black muzzle twitched up and down like a rabbit's nose. She began to lick the outside walls of the hut and, whether you believe me or not, that hut rocked to her licking. It was one of the most surprising things in my life.

I flapped some papers at her, shouting aggressively. She gave me a long wounded look and moved off to the corner of the hut. Now she began to rub herself, rhythmically, slowly and thoroughly: up, down; up, down; up, down. The hut swayed like a ship in a swell. She walked right round the hut, staring soulfully at me through all the three windows and rubbing herself hard on every corner as she went, her horns clattering loudly up and down the weather-boards. She licked the back of the hut, her rough tongue sounding like a carpenter's plane.

I put down my pen for a little while and enjoyed the lavishness of April green and blue before me. Coronet completed her round and started again at the beginning. Her head was already through the branch, and in a minute her shoulders would follow. Giving her a great despairing shove I gathered up my clutter and strode back down to the farmhouse.

Bill was sympathetic. He would put a fence right round my hut, he said; and he did.

Now, my only handicaps up there are a natural lack

of brain and a plethora of cats. The cats love the hut — I have had four in there at once — but Vashti is particularly taken with it, perhaps because she is in any case so taken with me. She sits on my shelf-table, edging always nearer in the hope of sitting on my page. She dabs at my pen, licks my nose and pats my notebooks to the floor; her hairs catch in my nib. But it is, of course, quite nice to be adored so long and consistently; especially by a Siamese cat for, unlike dogs, they do not give their affection indiscriminately. One can woo a Siamese for years without an atom of success; they are as unexpected in their loves as are humans. But they are faithful for life to the one whom they choose, and rough words from such a one can break a Siamese heart.

The hut is a place of refuge too from all the northerly winds. I have sat inside on many a March day — even in February once — with my sleeves rolled up and the sunshine warm and delicious, though outside the wind had still a winter bite in it. It is, too, a vantage point for quiet observation. From the hut one sees nothing spectacular: the distant figure of a hunting Siamese, a field's length away; a fox in the far corner; a pair of partridges out for a walk with their young; a falcon flying low past my doorway with a young bird in its talons; a rainbow round the sun; the tractor in the next field shuttling slowly back and forth beyond the hedge.

There is also the broader canvas of the field itself to watch from season to season and from year to year. No other field has been so much looked at, dreamed over and thought about, at least by me, as Lower Six Acres from my hut. Barn field is watched and discussed a great deal by all of us, for we can see it from where we sit at breakfast and lunch, at the long kitchen table; it is the view framed in Bill's new south window, with the corner of the barn for interest at the edge. But it is beyond both garden and yard and so not near enough for detailed observation.

Lower Six Acres comes right up to the front door of my hut, lapping the step with crinkling foam of bracken, for we have not yet eradicated the bracken from the borders of the field. Each crop we sow there is under my close scrutiny, as is each farming operation in the field — except, as so often happens, when I am working in the middle of it myself

Writers, I think, spend less time actually writing than they spend in thinking out their work. For hours, when I am thinking in the hut, my eyes are resting on the field. Consciously, I am not particularly aware of it, but its essence and character are soaking into my mind. For example: now, sitting at my desk in the farmhouse because it is raining and cold for June, I can call to mind exactly the way the field looked under the plough — the long brown ribs, straight but uneven, like lines drawn freehand by a draughtsman. Perhaps Bill has been square-ploughing and I can see the corners where the tractor and plough have turned, and the little central square where the furrows all run north and south, the whole looking like a small brown picture in the middle of a large brown frame. Or he has been ploughing along traditional lines with "lands" of perhaps ten furrows each in alternate directions and headlands round the margins, the "lands" showing up in broad stripes of light and dark brown, according to whether the plough was working towards or away from the hut.

I can remember the field harrowed, as it looked both from the hut and from the tractor seat when I was doing the harrowing; the jiggling clinking iron harrows smoothing out the furrows as a garden rake makes a tilth in dug-up borders.

I have watched the field as a pasture, with grazing cows and ponies; under standing corn or mowing grass; and lined with yellow stooks or green-gold haycocks. I have, of course, worked in it for hours, at all seasons and occupations, but I think I have seen more of it, more acutely, when sitting in the hut. From here one has time to watch the ways of grazing animals, to speculate on that one small area of field which they always avoid, and where plough horses, we are told, used to shy in sudden fright. Here in this corner we once left Coronet tethered, not thinking much of the rumours; but shortly afterwards we heard the sound of galloping hoofs on the high track by the yew tree. Racing out of doors we saw Coronet pounding along the skyline trailing her broken tether behind her. Right through the thick hedge by the walnut tree she plunged, down the slope of the Old Orchard and into the yard at full gallop where, treading on her tether, she flung herself to the ground with a horrible thud. Running to the yard we expected to find her badly injured, but she staggered to her feet, blowing and wild eyed but unhurt. We shut her into the cowshed for a while to calm down, but never again did we tether an animal in that corner.

What happened there? if anything? and when? Why do the animals fear that one place? It is a thing on which one may speculate endlessly, looking from the open-fronted hut. The grass grows there well enough, better perhaps than elsewhere for the animals neglect it. Flowers and birds are there, and we, with our dulled human senses, perceive no strangeness in it.

There is less excitement perhaps but more pleasure in considering the mowing grass and corn. The long grasses sway in the wind, and the wind moves over them stirring waves as liquid as a swell on sea-water. When it is a corn year, and the corn is oats as it generally is on this light land, the field is flaxen yellow when it waits for the binder. Oats do not have that sturdy red-gold, tan-on-rosy-cheeks look of the full-ripe wheat; they are more delicate altogether both in form and in colour. The shape of a single oat-head is like that of a little fountain, each individual grain-stalk drooping downwards on its thread-like arching stem. After a shower, when raindrops hang suspended from the grains, the fountain simile is still more apt.

The colour of a full-ripe oatfield is a little yellower than honey. As the day moves on towards dusk the yellow of the oatfield is intensified, as if it held sunlight, soaked up and stored until the evening. Looking at the oats from the hut on a July evening, all the light of the land seems to lie on this one plot; the field stands out from the rest of the farm as a sunlit patch does on a shadowed hill.

Rain dulls the colour of the oatfield; after days of rain it is like flaxen hair in need of washing and brushing, the sheen and life gone from it.

When the haying or the harvest is in full swing, and the field populated with its congregation of windrows or haycocks, sheaves or stooks, then I am usually in the midst of it and do not see it as it looks from the frame of my hut. I can hear the drowsy popping of gorse-pods in the hot sun, see the bold blue of viper's bugloss where the binder has trapped it in a sheaf, or brush the blue of harebells with my foot: I can feel the hot scratchy prickle of oat-stalks on my wrists and the stab of stubble round my ankles, but I am too busy to ponder over the picture as a whole.

When the corn has all gone, leaving the young grass ley that was sown with it for grazing, the field becomes a pasture again. It will remain a pasture for two years or so, until the grass is worn out and the circle starts again; plough, harrow, seed, harvest and pasture once more.

From the hut there is no sign of human habitation. There is only the quiet field, rounded like a pillow, the curled froth of bracken still persisting at its borders; the white hives droning with summer industry; the tall waving hedges and, beyond, the distant fields and a glimpse of the stony lane that leads to the Punch Bowl.

CHAPTER TWENTY

The Hammock

The hammock is like an eagle's eyrie; or so it seems to one so earth-bound — even valley-bound — as I.

Imagine Punch Bowl farmhouse, deep down in its little immediate valley with steep banks rising acutely to east and west of it. On the west bank — the buttress of Yew Tree field — grow two great trees; the yew, really a triple tree, vast and towering, sheltering the house from south-west gales; and an oak, like the yew taller by far than the house, this partly because of their situation on a bank and partly because both are fine and noble trees, full of years and lustiness.

The hammock, cunningly slung by Bill between the oak tree and a clump of hazels on the bank, hangs like a magic carpet swinging free in space. Higher than the top of the well-house roof, higher than the eaves of the farmhouse barely thirty feet away from it, level with the ridge tiles of the barn, it swings beneath its canopy of oak leaves. Lying in it (a thing I can rarely do, alas), I can see nothing of the ground immediately below it — the plateau near the top of the bank. But all around are the far-down fields and garden, seen from a kite's height; a bird's-eye view, a tilting large-scale property map rocking gently under my eyes as the hammock swings in its high branches. Almost all our near possessions are now lying in my sight, and I, in the hammock, feel as though I floated somewhere in the sky. I can see from the yew tree on my right, clear round by Barn field, the Little Orchard and the Old Orchard to the steepling pear tree on my left, to the house and garden set like a jewel in the middle and back to the oak tree branches like an awning over my head.

As the hammock tilts downwards the well-house roof lifts into sight, a pink and brown embroidery of old tiles and of clematis in bloom. Beyond it is the ancient barn, of stone and tile and timber, leaning with age and propped by two great tree trunks, its hipped roof surmounted by the black horse weather-vane fixed there by Bill and Robbie. There is a fair June breeze, good for haymaking (we shall soon be at it again) and the little horse swings between north-east and north-west. Beyond him, and actually behind him because of the steep climb of the further valleyside, is the long grass of Barn field, ripe for cutting. The whole of this field is now in flower, not only the white moon daisies, the sorrel, the buttercups and yarrow, but the grasses as well. The grass is in flower when the feathery heads are powdered with their pollen. This is the time for cutting, for when the grass has seeded there is too great a proportion of fibre to the minerals and protein, which in their turn have diminished.

In this lies the farmer's perpetual annual problem; to cut early and get better feeding value? or to cut late and get more bulk? Most farmers, I think, plump for the moderate course of cutting just at that perfect moment when the bulk and the feeding value complement each other best. And if at that moment the farmer also has good haying weather, why, then is one of the times when it is good to be a farmer. All too often in farming it is a case of "never the time and the place and the loved one all together."

Any day now, tomorrow perhaps, the sound of the mower will come from Barn field, but there are also Hanger field and Inner Wood down to hay this year and all must take their turn.

In Hanger field, when I am anxiously driving the tractor and hayrake over the plunging slopes of the hill, I shall probably comfort myself by calling to mind a remark of Mr Abels's when he was mowing for us once in that steep field: "I hear you once took a binder round here, but I feel as if I want a parachute."

In Barn field to-day the long grass lies under the wind and sun. The wind swings, but when it is blowing steadily for a while the grass becomes green water. Sun and wind together draw long ripples down the steep face of the field; silver ripples running down green water, for the leaning grasses throw off sunlight as the wind passes lightly over them. Never does any solid thing seem more completely turned to water than a meadow ripe for cutting or a cornfield when the corn is green, and a sunlit wind running over them.

Drifts of moon daisies swinging with the grasses now seem like fabulous pale seaweed swaying in the hurrying water. Only when the wind swirls does the field change character again, and then it is like blowing silver fire.

The perfect sound for a field of standing grass or

corn is that of the corncrake, like a massive watch being endlessly wound up. A drowsy June day sound, akin to the distant noise of mowing machines, I have not heard it since I was a child. Perhaps our modern mowers have driven it away, hurrying round the fields as they do and driving the birds in haste from their ancient nesting grounds.

Looking away from the hayfield I can see our Little Orchard, which we planted when we first came here. The grass around the trees is mown, brown hens and white hens walk about it and the young trees bow in the wind. There are small green cherries in the cherry trees; last year I dabbed the twigs with tree-grease in the hope of thwarting the blackbirds into leaving some for us, but they grew skilled at taking fruit while on the wing. This Little Orchard stands up like a wall above the roofs of cowshed, loose-box and stable, the small trees one above the other look like people in a grandstand. Through the half-open doorway of the stable I can see the round red rump of Clover, shut in because of the flies and her natural greed and stoutness.

The eye is now filled by the house and garden, the latter only visible, far below, as the hammock swings — now the south garden, now the north. The bearded irises are all in bloom, the exotic tangerine-coloured azalea Coccinea Speciosa, and three June flowering rhododendrons, purple, white and carmine. The Moonlight Broom is still in flower with the sun roses on the old wing site, and my new lawn is bright green, like a little curving carpet. The roses are late and have not yet begun, because of their removal in the winter, but I know the lovely evergreen azalea Satsuki is in bloom just out of sight, upon the far foot of the bank where my hammock is swinging. Through the oak leaves I can see the house roof, many tinted, moss encrusted, sweeping from the uneven ridge. Cheetah is howling softly to himself down by the doorstep, invisible unless I sit up, but I cannot do that for Vashti lies purring on my chest, delighted with the movement of the hammock.

Past the house to my left there is the tall Swing Tree, and young Ann Brown is swooping in and out of sight upon the swing.

The Pritchards' caravan is next in the panorama, waiting empty in the Old Orchard for the return of Robbie, Gail and Dion. Round it graze Coronet and the heifers Serenade and Tiara; Coronet alas is tethered again because twice yesterday she led the heifers through the hedge and down the lane. When the haying is over we shall have time to look to the boundaries again, but meanwhile Coronet's tether is long and she can reach both shade and water.

Now my view has come full circle, for behind me is only the hazel-crowned top of the bank with Yew Tree Orchard out of sight beyond it. All that is new is the sudden fleeting glimpse of Shelley's Cochin-China bantams, Blackberry and Blackcurrant, who live on this small plateau and can be seen by the hammock loafer if she leans out over the side, as from a boat. They are very shiny green-black little birds with fiery combs and curious feathered feet as if they carried fans. A more devoted couple one could hardly hope to meet, or birds more tame and friendly. And, as Shelley says, what does it matter if they rarely achieve an egg? When one is as beautiful as that, utility is not expected.

And now the sun has gone in, and Vashti has jumped down, and Ann is gone from the swing. The gate creaks and clanks at the end of the track, which must mean Shelley is home from school, and I must gather up my rug and go.

CHAPTER TWENTY-ONE

The Fox Cubs

On a cool June early morning Bill called to me in the house: "Would you like to see three fox cubs playing, in Lower Naps?"

Leaving a half-made bed I hurried downstairs and out of doors. Bill waited, a halter in his hand.

"I was just going to fetch Coronet," he said, "when I saw them."

We have often watched foxes and cubs around these fields, but we all still drop everything and go to watch them again when someone says that they are out.

Bill talked about the last evening's sunset, which I had missed because of having an early bath; one has to go into the higher fields to see a sunset here. The hills, Bill said, were all red-purple, the sky pure red like fire; he had never seen anything like that. I was looking at the young apple trees as we walked round the edge of Yew Tree Orchard, noticing that already they all badly needed hoeing again. There had been a lot of rain.

"The cubs are down in the bottom corner near the broom," Bill whispered as we came to the gateway. The grass in the field was thick and lush, starred with a Milky Way of blossoming white clover, full of bees. The three cows lay comfortably down in the hollow of the field; and Red Clover, grazing as usual because now she is so stout she has a heavy upkeep intake, lifted her head to look at us as we came in. We walked down the side of the high holly hedge, for cover, until the broomy corner came in view, and there were the cubs. They were about three months old, too young to be cautious. They didn't notice us, or if they did they were enjoying themselves too much to bother. The morning sun was behind us, making us difficult to see, and the wind conveniently towards us. Red Clover of all the creatures in the field stood watching us. She knew that any move from us in her direction might mean a morning's work, and the fatter she grows the better she gets at evasive action.

The cubs were exactly like any three half-grown puppies on any English lawn, except for their beautiful long thick brushes. Thinking of them in terms of puppies I decided they were just the type that would certainly have been docked: how strange a fox would look with the mutilated stump of spaniel, airedale or corgi! but no one thinks the dogs look strange; it is all a matter of custom.

We stood for five or ten minutes watching the fox cubs leaping and racing and chasing each other about, rolling each other over and pouncing with clumsy puppy-paws. One leaped at a butterfly, snapping its teeth; and another picked up a little dead broom branch, galloping round and round with it, pointed ears flat; tossing it into the air and pouncing on it where it fell.

"I must get Coronet," Bill said presently. "I've got a man coming about the lime at nine o'clock."

I stood where I was, now glancing at Bill walking along under the top hedge; now at the cubs which, still wildly playing, took no notice though Bill was in full view of them and moving. Coronet stood up slowly, arching her tail as she stretched herself. She was like a figure on a shield, her smooth back dipping to her stretch, her muzzle tucked in. Walking up to her Bill slipped the halter on her head and began to lead her up the slope of the field, her hoofs and his shoes hidden in the dew-wet clover.

The cubs stood a moment looking at them, for this was too obvious a happening to be entirely ignored; but in a minute they were playing again. Now Clover suddenly shook her head and galloped up the field after Coronet. Serenade and Tiara slowly followed; they were both heavy in calf. Again the cubs looked, with sharp ears cocked, but again they dismissed the affair as not worth bothering about and went on playing. They were extremely pretty, with their warm rich body colour, pale fronts and white tail-tips. Only the male has a white tail-tip, so these three were all dog-fox cubs.

Bill and the cows and pony were gone from the field, and I stood where I was for some five minutes longer, till one cub suddenly raced away down between the yellow broom clumps to the wild valley. The other two played for a little longer and then they too raced away to the valley, a jay screeching as they rushed past his tree. There was no one left now in Lower Naps but myself and the bees in the clover. I turned and went up the hill into the sun and back to the farmhouse.

CHAPTER TWENTY-TWO

Silver Wedding

Early in August Kenneth and Eva were to celebrate their Silver Wedding. We were delighted at the farm to learn of their hopes of having the celebration here, and said so at once, though with some doubts about the suitability of the place. They were not put out, however, by our suggestion that twenty-five people making merry on the ancient sitting-room floor would almost certainly be cast into the cellar beneath, by the sheer weight of their numbers. What could be nicer, they said, than a farmhouse kitchen tea-party?

This idea was so obviously the right one for our farmhouse and our kitchen that it was carried at once. In my mind's eye I saw the kitchen, so recently dark as a cave, more bright and festive than it had ever been before. Sunlight through the old and new windows would gleam on white icing and little silver slippers. Punch Bowl Farm would have its great bowl of punch for the drinking of toasts. Its bowl of whipped cream would be adequate for all the Devonshire splits and fruit melbas that a kitchenful of people could eat.

The Webbs and young Pritchards arrived with their two caravans, and, in a whirl of farm work, holiday exploits, and the celebrations of both Dions' birthdays, the wedding theme gathered strength. Bill produced a cob-webby trestle table and benches from somewhere in his bound-to-come-in-useful hoard, and the table (well scrubbed) was pronounced just right for a serving-table by the caterers who came out from Godalming to consider the amenities. (What would they have made of a hundred-foot well and no drainage?) An elegant large gate, bought at a sale and given by Bill's brother Sam, was painted white by Robbie and Old Dion, newly lettered by Eva and hung in place at the farm entry by nearly all of us. Someone remarked that only a lodge was now needed.

In the garden there was a sudden surge of willing labour, tidying and mowing; and on the still chaotic parts of the old wing site the clearance was such as would have taken me weeks of solitary toil. The list of guests was fitted to the capacity of the kitchen and the seating planned. The invitations were accepted by people living at what seemed quite forbidding distances. The original best man, himself, proposed to drive his family up from Exmouth, and back again the same night, and nearly everyone else would be coming from outside the county.

We thought about the flower decorations. The caterers had offered to do these but Eva, fearing the usual plated vases of pink and white carnations, suggested that we do them by ourselves. The evening before the day we went up into the high fields to find such things as were both decorative and essentially of the farm. Great armfuls of tall willow herb and silver birch branches would make a background for large-scale arrangements. We picked stems of oats and barley, green and gold bracken, the earliest heather in bloom; and coming home through the half-cleared Upper Six Acres we stopped to pick a few sprays of potato flowers from the patch Bill and I had planted by the beehives in the spring.

"I believe we could pull quite a lot of legs with these," said Eva, smiling.

"They do look rather attractive," I said. "Strange and faintly subtropical — if you didn't know already what they were."

The flowers were left in the cool cellar in jugs and buckets all that night, and the next morning Eva took her scissors and all the flower containers the house could muster and went down into the cellar to do the arrangements. She says she now knows how to climb many times up steep steps, open a door towards herself and fend off four cats without dropping anything.

The weather was quite wonderful on the day. We were tempted to have the whole party on the lawn, but in the end decided not to make ourselves hostages to the British climate. So we put out all the deck-chairs, garden chairs and rugs that we could find, and were thankful that we could spread our numbers comfortably in the sunshine all the time except during tea. Everyone arrived without mishap, including Kenneth's father (in his eighties) who made the journey from Croydon, and the Exmouth party who were here in good time. We sadly missed Vera and Norman, who were abroad at the time, but Robbie, Gail and Dion were here to represent the Pritchard family.

If you have a party at a farmhouse in fine weather

you must expect it to be a mobile party, especially if nearly fifty per cent of it is in its 'teens. Ours was a farm party with time to see the farm. There was also, of course, a great deal of interest shown in the house, and, while some were arriving and being introduced, others were ducking through the four-foot bathroom door, negotiating the cellar steps, admiring the flower-decked inglenooks and beams and new windows, and making reticent remarks about the potato blossom.

Back in the sunshine out of doors, the party scattered, but soon collected at the call to tea. The kitchen was so crowded with chairs and benches and tables, with flowers and food, kettles, urns and crockery, that I felt I hardly knew it. I thought how I should have liked all the past farmers and their families, who had lived in this old house, to have been able to look in and see it as it was on that day — so festive and flowery and gay, so full of friendly people.

Each place was marked with a labelled posy, and the guests all fitted in like pieces in a puzzle. I feared for the cats, among so many feet, but they were astonishingly unconcerned; two strolled out, one found my knee and another someone else's.

The caterers were efficient and unflurried; one would have thought they did most of their work in farmhouse kitchens. The Aga was entirely covered with hissing kettles and the doors and windows stood open to the August afternoon. Young people in the country have amazing appetites and soon the piled tables began to look less burdened. Presently even the cake was only a half-moon. Now the punch was ladled out for toasts.

There were a few short speeches, Kenneth paying a delightful tribute to his years with Eva and adding at the end, "Eva and I would like to drink a toast with our sons."

There was a moment's abashed silence in which Dion and Julian held their glasses upside down, but their dilemma was quickly seen and dealt with.

After tea there was a farm walk, in continuing glorious sunshine. The company fared forth in two parties, the young and the not-so-young. It was pleasant to be able to go with them, knowing that the caterers were clearing away and washing-up, and that there was abundance left from tea for a cold snack supper.

The ladies had all been warned to bring spare shoes, and they soon saw the reason for this when we left the fairly level path beside the pale gold oats in Lower Six Acres and arrived at the scene of the clearance work in Upper Six. Here the old turf tracks were cut across by sandy furrows, and the unploughed areas were spiky with burnt brambles, studded with tree-stumps and other traps for the unwary. No one minded. Kenneth's father was resting in the garden after the excitement of the afternoon: the rest pushed gamely on and were soon out in the wilds of Nameless field, peering over the tall waving bracken and through the jungly hedges to the jungly land beyond.

Passing over the farm boundary we went on through Highcomb Bottom to look into the Devil's Punch Bowl on a perfect summer evening, where those with any sense of time left began firmly to steer the party homeward: back through Nameless field and Upper Naps with the blue distances northward, Lower Naps with sweeping curves lapped round by woods, and Yew Tree field, level and tamed, waiting through years for its harvest of apples.

So soon the party was over; there were trains to be caught; cars moved down the track and through our handsome new gate while hands waved and goodbyes were shouted. The young folk vanished a while, reappearing quickly in farm clothes ready for rough games with Bill, for milking and pig-feeding, and for bonfire building in Upper Naps.

The bonfire was the crown of the day's celebrations. We knew it was to be a good one but we hadn't expected the tower of branches that we found there when we walked up through the fields at dusk. Looking at it, I wondered if we should have phoned the Jupes and the Abels's, so that they would not think we were on fire, but the undergrowth was still wet after several rainy days and I decided that the blaze would not be great.

Everyone was up there in the dark field, including Shaun who stayed up late especially. The clear sky over us sparkled with stars as Bill lit the first small earth-star that was so swiftly to grow into our bonfire. Brightness and warmth suddenly encompassed us: this fire was no chicken-hearted glimmer discouraged by the prevailing dampness; it was a success, a fire to remember, a beacon — a beacon, I thought a little anxiously, recollecting our neighbours on the hills to east and west of us. But no one else worried. It was not possible to worry for long, with all this leaping, exulting fire, these showers of coloured sparks like a rain of rubies high in the air, this warmth sinking into chilled bodies. Someone dragged along to the sacrifice the ruins of my old red hessian loose-covers (what else can you do with old hessian, relic of war-time, anyway?) but we, the middle-aged ones, abstracted them for our immediate comfort. The fire roared and swirled up higher, and we admired it from our couch of old red covers on deep heather.

Black figures moved in ritual attendance on the blaze until the heat was too intense for anyone to approach it. Now the flames were more than thirty feet high, showering glowing sparks like a tall coloured fountain. All round Upper Naps the trees were lit up, strangely beautiful in the eerie quality of firelight. Like stage-trees they were, unnaturally green, brilliantly lit, dramatically shadowed.

When the red heart of the fire glowed like a little sun, Bill took up a bag of large potatoes. Borrowing from us — the lazy watchers — a hessian cover, he draped himself inside it like a monk in a lurid red cowl and approached backwards to the fire, dropping the potatoes into the hot ash. It might have been years and years B.C., and we a band of Druids, and the night a night of sacrifice.

At midnight Shaun and I went back to the farmhouse, leaving the others to wait for their roast potatoes and make their dough twists. As we walked into the kitchen the telephone rang. It was Mrs Jupe. She had been trying to get through to us for the past half-hour, she said, as they had feared we were on fire and possibly in need of help. They would gladly have come over.

With apologies and thanks I explained the circumstances. "I had been thinking that we should really have rung you," I said, "but we never expected such a blaze, ourselves."

"The sky is simply red, from here," said Mrs Jupe.

It was now well past midnight, but any hopes I may have had for getting Shaun quickly to bed were dispelled by the sudden arrival in the kitchen, via an open window, of a fine large tiger moth. At once there was a commotion, the cats all competing with Shaun to catch the moth.

"My net!" said Shaun, dashing through the house to his room, then back again with his butterfly net. The cats were beaten to it, the moth caught and killed (I am afraid I always look away), and then carefully fixed on the setting-board.

"Now, come along, old chap — it's already tomorrow morning."

"Oh, but Mother, what about supper? Well, I know I've had one, of course, but that was hours and hours ago. I'm simply famished. And, after all, it was a silver wedding."

Much later, the other young folk came tiptoeing quietly into the house. Tea? Early supper? Roast potatoes? Dough twists? Oh, yes, they had had all those; but what they wanted now was something to eat! Famished? Well, if I didn't believe them I could just watch what they could do with any food I had around! The time being what it was, it seemed best to believe them.

CHAPTER TWENTY-THREE

Milking Parlour & the Beetle

In the same August we made a decision which now, at the time of writing, engulfs us in a wave of extra work leaving no time or energy for any but the barely essential jobs about the house and farm. We decided, in fact, after all these six years, to increase our cattle and sell milk. We might never have made this — to us — bold and precarious decision had it not been for the birth in that month of two bull calves — to Serenade and Tiara. These were our third and fourth bull calves in succession, our last heifer calves being Serenade and Tiara themselves, two years ago. With no heifer calves to rear, the house was suddenly flooded with all the milk from these two freshly-calved heifers, as well as a small amount from Coronet, she being near the end of her lactation.

We had two alternative policies to consider: buying heifer calves to rear (or even bull calves of a beef breed, but this is a heart-breaking business); or buying older cattle to bulk out the milk for selling commercially. After much deliberation, and bearing in mind the quicker profits from milk, we decided to buy older cows, with young stock to follow on, providing that our old buildings were passed as suitable for selling milk.

Bill didn't think the buildings had much chance. They were incredibly old, small, and dark, and were built of Bargate stone and timber with old tiled roofs. The cowshed, however, had a concrete floor and a fairly modern drainage system. There was also a concrete approach which Bill and the boys had made during the school holidays.

For my part, I never seriously doubted that we would be allowed to sell milk from the buildings. I knew we couldn't hope to get an attested licence, of course, despite the fact that the cattle themselves were all attested; because for this one's buildings would have to conform to a very rigid standard, both as to construction, drainage, approach, and measurements of standing room per cow. Besides which, one would need a satisfactory dairy for cooling milk and for washing and sterilizing equipment and utensils. But as to selling milk for pasteurization — farmers had always done so from this farm and from these buildings. In fact, milk was sent from here right up to the time of our purchase of the farm. Besides, I said, what about so-and-so's farm? He was certainly selling milk — he always had done so — and his buildings were at least as old and unorthodox as ours were.

"I dare say," said Bill, "but once a licence has been withdrawn from a farm it may be difficult to get it renewed, without coming into line with new standards. The fact is, we haven't a licence to sell milk, and I don't think we shall get one, with these buildings."

I said I didn't think he could possibly be right,

but we applied to the Surrey Agricultural Executive Committee for an inspection. Meanwhile, the household both enjoyed and gave away great quantities of cream and butter, and Blackie, Dingle, Bellybody and Cuss fattened on the buckets of skim milk.

The A.E.C. Milk Production Officer was, I am happy to say, a woman; one Miss Clare. I am feminist enough to take delight in seeing my sex ably holding traditionally male professions. In fact, so ably did this one hold her job that I soon saw (with mixed feelings of admiration and sorrow) how fatuous were my hopes of getting her to pass our ancient buildings. With measuring rod in hand, Miss Clare demolished my ambitions. The cowshed was too narrow. The whole front wall would have to be rebuilt three feet further away and the roof brought out to meet it — in any case, she added, it would have had to come down as it was largely timber, and it must be brick or breeze block, nine inches thick. All the timber must come out, she said; the wooden partitions and posts and the end dividing wall, and we must have a concrete "apron" outside.

"Even for selling ordinary milk?" I asked. "Without an attested licence?"

There was no longer any difference in the standards, said Miss Clare to my astonishment (the 1949 Milk and Dairies Act having passed right over my head). These were now the requirements for the production and sale of all milk, from all cows, the attested licence depending only on the attestation of the cattle.

Oh well, said Bill philosophically, at least if we did do the job we should know we were getting the maximum price for what would be, after all, the highest quality milk.

"But it does seem a pity," I said. "I mean, although the buildings are old and difficult to keep clean, we know that they are clean, and that the milk is pure and cleanly handled."

"That may well be so, in your case," said Miss Clare. "The trouble is, we can't make special rules for special cases. Unless we do insist on certain standards in buildings and equipment we can never be sure about the handling of the milk in all cases."

She looked over the half-door into the calf-box. "You'll have to have a dairy, of course, and that, too, must have a concrete floor, brick or breeze walls, and a ceiling under the roof. This loose-box could very well be adapted, though you'd have two new walls to build: these timber ones would have to come out."

Suddenly I had an idea.

"The old stable is much wider than the cowshed. We hardly ever use it. If we could adapt that it would save bringing the cowshed roof out."

We went down the yard to the stable, Miss Clare looking so completely feminine and immaculate (we had expected somebody in gumboots) in well-cut suit, sheer stockings and faultless shoes, that her efficiency would have been hard to believe in, had we not been experiencing it for ourselves. For some while we all compared and discussed measurements, this way and that, but though the stable was wide enough from head to tail of the cows, it was too short in the other direction to house more than six cows side by side — and those squeezed in to the final half-inch of the measurements required per cow.

"What you really ought to do," suggested Miss Clare, folding her rule, "is to make this building into a milking parlour and dairy. It would be far the least expensive thing to do, because you need only have two standings, and the dairy would be under the same roof."

Now we had, ourselves, considered this before, but our objection was that two different buildings would have to be in use for the cows at the same time. A milking parlour (the correct term, strangely enough: you can call it a factory, but hardly anybody does) is never used for anything but actual milking. The cows do not lie in it at night, nor are they kept in it at all except while being milked, which they can be two at a time. Consequently, the measurements allowed per cow may be much smaller, the drainage and ventilation simpler, and complete isolation from the dairy is not necessary, as it would be for an ordinary cowshed. The most serious drawback, to a farm with so few buildings as ours has, is that a second shed must be used for lying in on winter nights, unless the rather uncertain plan of lying out in the fields all winter is adopted. (What if there is a very hard winter?) Also, there is a certain amount of extra work in fetching in and turning out the cows, two or four at a time, for milking.

When Miss Clare had departed we talked the whole thing over. First, were we really going to tackle the buildings at all? Or should we drop the whole scheme, buy in a couple of calves to rear and forget that we had ever thought of starting a herd? We leaned very much, now that we had entertained the idea, towards having a go at the buildings. But adapting old farm buildings can be a fabulously expensive thing. The most humdrum, utilitarian modern cowshed can cost as much as three thousand pounds: a milking parlour and dairy can cost quite half as much. How many years would it be before so much capital outlay was repaid in milk from the farm?

"We could do it all ourselves, I dare say," said Bill reflectively. I wholly believed him. After installing the bath, w.c., central heating and drainage in the farmhouse, what was there to a milking parlour that could defeat him? I knew, of course, that I would be conscripted as builder's labourer, and would surely be running up and down ladders with tiles and nails and whatnot, but I wouldn't have to work the whole thing out. Besides, we hadn't decided to do the buildings, yet.

It was our fate, just at about that time, to hear of a Jersey cattle farmer near Farnham who, being overstocked, had decided to sell a few head of cattle. We telephoned our inquiries. Mr Marwick, the farmer, answered them, listing the beasts he had for sale: there were seven altogether, of varying ages, all pedigree and attested. The reason for selling seemed sound ("We've had a run of heifer calves — thirteen out of the last nineteen"), and we liked our impressions of Mr Marwick himself. We decided to go out and see the cattle.

This trip was really rather funny — except to ourselves at the time. It happened during a period when we had no car (after the demise of our ancient Standard, Thunder, and before the purchase of a slightly less ancient Hillman Minx, Delilah), so we borrowed, by the kindness of Mary

Fisher, her little three-wheeler — known as The Beetle by Mary, and The Bumper Car by Bill. This little item has a horse-power of two, is driven by a chain, like a tricycle, and has its odd wheel in front instead of, as more commonly, behind. (We were to regret this detail later on.) But meanwhile it carried Bill and Shaun and me to Redfields Farm in creditable style. Shelley remained at home to cook the lunch, and because two and a half persons is the full capacity load of The Beetle.

"Don't be late," she said, "because of spoiling my lunch, and besides, the Jupes are coming to tea."

I said we ought easily to be back by one — in fact, we must be. Cramming aboard with Shaun in the middle, we set off with a noise exactly like the Allen Motor Scythe (naturally, the engines being identical). We stopped in Elstead for some petrol and managed to start again (after I had pushed at the back for a bit); and, gathering enormous confidence in The Beetle, we drove right up to the door of Redfields Farm — a distance of about thirteen miles — without a single further incident.

Our telephone impressions of the Marwicks were in no way lowered by meeting them: we liked the whole family — Mr and Mrs and three children — and began to feel more confident about the selling of the cattle. You simply cannot be too careful about the purchase of any livestock. In practice, almost no good animal is ever offered reasonably for sale. No farmer is going to sell a profitable animal without very serious cause, and genuine overstocking through runs of heifer calves is an unfortunately rare phenomenon. Mr Marwick, however, apart from just not looking the type who would "do" you, was prepared to provide certificates of good health with all his animals. All the seven were maiden heifers (which means that none had yet calved), so certainly no fault could exist in milking or fertility. The first one of the seven to calve would be Redfields Misty, about a month from the date of our visit. None of the others (ranging in age from five to fifteen months) had yet been stocked.

We liked the look of the cattle, too. We didn't particularly want animals in milk, in a hurry, until we had had time to get the buildings in order, and most of these seven would calve down in the following late summer and autumn. Thinking of January to March, however, when milk prices are at their highest, we did inquire about possible milkers later on. The Marwicks mentally considered their milking herd — and later we all went out to look at them — but they decided that none could be spared. It was as well, I suppose, as it gave us more time to get our place in order; and we had been offered a grand old milking cow from the Jupes' Ridgeway herd which we would probably accept if we bought these Redfields seven.

Bill had been making notes about the heifers for sale as we viewed them, and these we checked in Redfields farmhouse over a welcome glass of cider, while Shaun and the two youngest Marwicks played with a railway laid out on the floor. There were three young calves, Heatherbell, Venus and Maidenhair, aged five to six months; Sapphire, a lovely but shy little heifer of eleven months; Ruby and Dewdrop, exuberantly friendly and about fifteen months old; and Misty, nearly two years old. The prices seemed to us reasonable and, cautious though we were, we could discover no snag. We said that we would think it over and telephone later.

Bill actually started The Beetle without my pushing, and we departed with a flourish down the drive.

All went well until we were just entering Farnham, down Castle Hill, when there was a most appalling grating noise, with violent bumpings and rockings. Shaun and I clasped one another as Bill stepped on the brake. The moment I could, I scrambled out. The near-side back wheel was a ruin — inner tube bulging through the tyre, rims bent, paint scratched — after ten yards of too-close contact with the kerb.

"It's being so much wider at the back than at the front that's so deceiving," said Bill (I showed sufficient self-restraint not to say, "But I kept telling you to keep out further from the kerb"). "Oh well, we've got a spare wheel," he added cheerfully. "Pity, though; it might make us a bit late, and I'll have that bent wheel to do, some time."

"I thought it was rather a demented design," I said, "the single wheel at the front. The driver can't see the widest part. However . . . can I help?"

Bill said he thought I'd be most help if I went for a little walk with Shaun. Taking his jacket off he sat down on the kerb and addressed the wheel. Shaun and I went up the shallow cobbled steps to look at Farnham Castle. It was our luck that, being a Sunday, the gate was not open to visitors till the early afternoon, but we enjoyed the peaceful sunshine in the courtyard where no sigh of wind passed to stir the flowers below the walls. I thought what a wonderful place it would be for writing (except for the visitors), with the sun pouring down and the tall walls keeping out the wind.

When we got back, the new wheel was still not fitted, so we said we would walk on down Castle Street into the town. Shaun was looking for caterpillars and chrysalids in people's hedges, and I was looking for favourite shrubs and trees in the gardens beyond. Bill caught us up at the bottom of the hill and we thankfully climbed in again. The Beetle stuttered on, in her loud fashion, for exactly the length of one short street and then drew sadly and weakly to a stop.

"Gosh," said Bill, "I believe the chain's come off."

"It can't have," I said fatuously.

Bill stepped out and made his diagnosis. "It has though; and, what's more, it's gone."

"Gone?"

"Lost," said Bill calmly. "I expect we've dropped it on the road. It can't be far back; you stay here and I'll go and have a look."

Now actively hating The Beetle, Shaun and I walked up and down as if we had nothing to do with it. Mercifully the sun was warm and the September day very pleasant, for it was a long time before Bill reappeared. We had meanwhile been staring into the windows of a handy garage, discussing which car we would have if we could choose.

"I've got the chain," Bill said, "but there's one link missing, and I couldn't find it anywhere."

"Oh. Perhaps a garage might be open, somewhere, and have one," I suggested.

"They aren't very easy to get," Bill said ominously, "except from cycle shops. Anyway, I'll see what I can find out."

This time Shaun went with him and I sat uselessly on a public seat, surrounded by about sixteen dozen scarlet geraniums and one terrified pink one. Time passed. Ultimately, though I had given up nearly all hope of ever seeing either of them again, they reappeared beyond the geraniums and broke the solemn news to me at once. Not a link to be found in all Farnham. Of course, it was a Sunday.

I said, practical even in despair, "I'd better telephone Shelley. It's already after one o'clock."

"There's a phone box down the road, just by the fire station," Shaun said and took me along to it. Bill went off on other projects, the wretched oily chain curled up like a viper in his hand.

Shelley did not sound surprised. In fact, she said she was not — with The Beetle. "Then I'd better have my lunch," she said, "and save some for you. I've made rather a nice first course, and after that there's a trifle." She said she would ring up Ridgeway Farm and explain matters to Mrs Jupe, and then expect us some time before nightfall.

"We'll probably have to get back as best we can by buses and walking," I said. "Do leave plenty of lunch, won't you? It seems weeks, already, since breakfast." It was only much later on that I discovered Bill hadn't had any breakfast at all that day, what with one thing and another.

Shaun and Shelley had a private conversation for

some minutes longer while I crept back to the public seat and the geraniums (some poor gardener must have been in trouble over that one pink one, I thought).

Time passed. I supposed my youngest dear child must by now have linked up with Bill; in any case he knew all too well where I was to be found. At my time of life, much standing is in no way called for.

It must have been about two o'clock when they reappeared, triumphant.

"The chaps at the fire station have a three-wheeler, and they've got a spare link. It doesn't fit our chain, but they say they think they can make it fit."

We all repaired to the fire station, where everyone was quite charming and anxious to help. I noted with feeling that their three-wheeler had its single wheel at the back, and pointed this out to Bill. But the affair of the wheel was now such old history to him that he was not disposed to call it to mind. The present was what mattered. Another half hour passed, in which Shaun had a wonderful time around the fire station, and the fire crew actually got their link to fit our chain.

"Now, I suppose we really can get going," I said hopefully, after thanks had been duly pressed where so patently deserved.

"Well," said Bill, "I've got to fix the chain back on first, of course. We've parked The Bumper Car near a church, down a side street. It shouldn't take more than half an hour."

I left him, completely under The Beetle now, in all his Sunday finery, while I went to see if a single food shop might be open on a Sunday in those parts. There may have been one, but if there was I didn't find it. When I came back all was as before, only both Bill and The Beetle were now so dirty that I didn't much like to seem connected with either of them. I sat on the churchyard wall, which had a peak and was therefore most uncomfortable for sitting; and in any case I had to get up every now and then to make room for Shaun and two urchins he had collected who were all having enjoyable chasing games along the ridge. Time passed, only enlivened by Shaun or I having to get inside The Beetle and pull a lever every now and then, which in some way helped Bill to get the chain on.

We did, in fact, get away from Farnham that day, and in The Beetle. A further remarkable thing is that we stopped for at least twenty minutes on the way home at a large pond called The Moat, where Bill and Shaun desired to look at some model yacht racing.

"Aren't either of you ever hungry?" I demanded, after this time. "And to think that you haven't had a thing to eat, to-day!" I said to Bill. There is no understanding some people.

All right, they said, they were coming. They had just forgotten about food because of the yachts being so interesting, but really they were very hungry indeed.

We got home a little late for tea. Bill made one meal do for the lot, and then went milking.

CHAPTER TWENTY-FOUR

Punch Bowl Herd & a Transformation

That evening we worked out costs, then we worked out labour (only Bill and I with Shelley at week-ends), winter feed available on the farm (hay, feeding straw, oats, barley, kale), and probable ultimate profits. We drew plans of milking parlours, cowsheds and dairies, for comparison of cost and labour. Bill said that he could do all the practical work of converting the buildings by himself, or with very little hired labour: doubtless I wouldn't mind myself doing easy little jobs such as retiling the stable roof. No, I said, of course not; having long prepared myself for the idea.

There were already bricks and tiles stored from the old pigsties which I had years ago demolished for the sake of the view through the new kitchen window. Cement, sand and ballast would have to be bought, as well as hopper windows, dairy fitments and so forth.

So we talked, and then we stared at our papers for some time.

"Well, shall I ring Mr Marwick?" said Bill.

"I think so," I said. "It's a good opportunity to get

decent young stock, reasonably, and from the breeder. Besides, I suppose we'll have to start farming properly some time. It may as well be now, while there is anything at all in the bank to cover it, and while we're still young enough to face the initial toil and struggle fairly cheerfully."

Before going to bed that night we had bought Heatherbell, Venus, Maidenhair, Sapphire, Dewdrop, Ruby and Misty; and we had decided to follow Miss Clare's idea of turning the old stable into a milking parlour and dairy. Further than this, we also decided to buy an eighth cow, Golden Goblet, which was the good old retainer already offered us by the Jupes. At eleven she is pretty old for a dairy cow, but she is a lovely type of Jersey and has calved regularly every year of her life since she was two years old. She was so cheap that we could hardly have lost on her; and if we succeed in getting her in calf again even once more she will have paid her way quite comfortably.

The cattle arrived before we had organized our preliminary attack on the stable. It didn't matter very much, for only Goblet was in milk. The Redfields heifers came in two instalments in Mr Marwick's double horse-box towed by his Land Rover. We put the calves in the empty calf-box, where they had an electric fenced run in the farmyard at their door. The older four were driven up to Lower Naps, away from our own three milkers who were then in Hanger field, now full of good grazing after hay.

Golden Goblet came on her own feet (our boundaries marching with the Jupes'), carrying her tail over her back in a way quite characteristic of her and, Mr Jupe tells me, of most of her calves also. She settled down at once with the milkers in her sedate and rather distant way, keeping herself to herself as she used to do at Ridgeway, making no fuss and accepting all things as she found them.

The autumn term began and Shaun went back to school — this time in Delilah: it was her maiden voyage, as one might say, since we bought her.

Now, the long summer holidays being over, Bill and I let everything else slide and started on the milking parlour and dairy. For a while we lived mainly on old hens. These, moulting now and past their best laying days, made three successive dinners each: first day, roasted for half an hour after slow overnight cooking in the bottom oven of the Aga; second day, cold with salad and jacket potatoes; third day, curried, or stewed with dumplings. Old hens are so tender, after this long slow cooking, that they can easily be eaten with spoons (I think Bill would have done this, had I not been there, so that he could draw plans or tot up measurements without paying attention to the dinner).

My first morning on the stable roof was a pathetic affair. Never caring for heights, I scaled the home-made ladder slowly and deliberately, not looking down, never leaving go with both hands at once, and carrying with me one nail and one tile at a time. Bill, who had fixed my ladder in place against a straw-sack buffer (the buffer was to protect the roof, not me), now hopped about the bare rafters like a canary, nailing new battens in measured rows to take the tiles.

We had already done some patching on this roof, in

our earlier years, so now had only about half the roof to strip, rebatten and retile. My initial uneasiness soon wore off and I was able to write to my mother that night: "I am getting very nimble on high places, dashing up and down the ladder with eight tiles and a tin of nails and half the time not holding on."

Whenever I peered down through the cobwebby rafters into the dark stable underneath, however, the old wave of giddiness swept over me. I also felt faintly sick when I looked for long at Bill's sanguinary passage across the gaps with his hammer, nails and armful of new battens. My own job, once clear of the fiddling whole-tile-half-tile edge pattern, was delightful — like building with a box of model bricks. Each tile has a pair of nail holes in the top, and all I had to do was drop a large galvanized nail into the left- or right-hand hole, as required by positions of intervening rafters, and hang the tile by its nail upon the batten being covered. The next row is placed half a tile further along, so that no gaps are left for rain to pour through.

We were using the same old tiles again, of course, except for a few broken or perished ones (tiles do perish, in time — a thing that surprised me when I heard it), and they went back on to the roof complete with humps of vivid moss, golden lichen patterns, and all kinds of insect life, most notably spiders, earwigs and woodlice. At first I used carefully to brush off all this wriggling and rushing populace, being very touchy about that kind of thing; but after a time I realized, with a surprised little shock, that I was tucking the whole congregation under my arm with the tiles and never sparing them a thought.

So, in the stresses of emergency, do our most rooted feelings sink into the background.

It was pleasant being in a position to view so much of the farm at one sweeping glance. We were able easily to track down a wandering hen who had ambitions to lay her eggs in the thicket at the back of the cowshed; we could see what all the milking cows were doing in Barn field, and when the new heifers came down from Lower Naps for water. Anyone passing up the drive to the house was spotted at once.

Freyni was the only cat who took any interest in the tiling, and she took too much. Whenever I was working on a most complicated bit (such as where a row didn't work out properly to the eaves, the tiles being irregular in shape and size) she would suddenly appear, rush lightly up to some inaccessible perch and there cry out most piteously that she could not get down. I would have to extend to her a length of heavy planking, holding it as firmly as I could in straining arms while she dithered down it to the ground. This happened so often that, had I not come nearly to the end of the tiling, I would have hardened my heart against her.

She always was a cat for creating dramatic situations around herself. An old trick of hers is to hold up one paw, as if wounded, and thereby attract a lot of sympathetic attention. Once, when she really did have an abscess forming in a hidden bite on a foreleg, nobody believed her till she had been holding it up for nearly two days. She also practically obliges other cats to chase her, by rushing pointedly away from them, emitting provocative little trills and twitching her ears. She then tears up into the bird cherry, looking pitiful and persecuted, in the hope that someone will climb up and get her down. No one does. We all know that she can go either up or down that tree with equal ease, providing no one is openly watching her.

Freyni, by the way, is now a neuter (spayed, as the term is), the same as Vashti, after having a difficult kittening in the spring. This had ended in a Caesarean operation and the loss of all her kittens. Poor Freyni! She who had always loved all kittens is now as crusty an old aunt as ever Catto could be. Both of them, at this time, were having psychological trouble together, and the cause, of course, was a new lot of little strangers recently born to Thistledown and Cheetah.

The Punch Bowl cats were now divided into two distinct camps — pro- and anti-kitten. Though Cheetah at first assumed a detached attitude to his four squawling atoms, as they began to develop a little character he (male-like) became more and more fascinated. By the time they were a month old, and promoted from the copper to the basket, Cheetah idolized his children, defending them hotly from the aunts, with fur flying. All night and half the day this big bumbling tom-cat would spend curled up, blinking sentimentally, around his fighting four. He washed them regularly and lovingly (though, male-like again, a couple of licks each were his idea), he let them bang him and whack him and chew his silvery ears. The only thing he really failed in was the rations. He just didn't have any; and being a far better parent than Thistledown, he was oftener with the kittens. I think this worried him quite a lot. He used to sit there in the middle of them while they screamed for sustenance, his ears slowly twitching and his blue eyes growing rounder and rounder. Then suddenly he would rush out from the basket and sit on the rug and meditate.

Now that they are a little older, and on a mixed diet, he really can and does enjoy them. Never was tom more homely, more devoted or more proud. I only wish that his family feelings did not provoke such growing hostility with the aunts.

Bill, having finished nailing battens, left me to complete the tiling while he began on the work below. In less than a day he had stripped down half the weatherboarded front wall to its upright joists, torn down the rotting hayrack (which made lovely kindling wood for the sitting-room inglenook) and removed the old wooden partitions. I could peer through the grill of battens and see him at it as I worked on the roof.

When he had cleared the space around him, Bill began to pull up the old broken and pitted cobbled floor, levelling and packing the undersurface ready for concrete. I finished the roof and left him to it for a time while I caught up with my writing. From this point we employed a little outside labour; Bill Davy, the cowman at Upper Highfield, obligingly came along for a couple of hours in the evenings after his own day's work was done. The stable, brilliantly lit by a one-hundred-and-fifty-watt electric light, was under my eye as I worked at the window of Shaun's little room — the easiest place to heat when we were both too busy to fell and saw logs for the big inglenook. It was a strange sight, rather like a stage set, with the front wall almost open to the night and the inside lit up like a backcloth against which the two men bent and moved, laying the concrete floor in boarded panels.

Instead of scoring diamond patterns in the unset surface — the old anti-slip device — Bill scattered carborundum powder. This, setting hard in the concrete, makes a good non-slip surface which is far easier to clean than a scored surface, and is also quite attractive to look at, the carborundum catching the light and sparkling to it.

As soon as the floor was finished the front wall began to go up, with a new window set in it to light the dairy end of the building. At the time of writing, the stable looks like a hybrid, jester-coloured place — half red brick and half the old silver weatherboarding with which we had become so familiar. It reminds me of a semi-detached house where no unison of ideas exists about the outside decoration: sometimes one sees such a place where individuality and non-co-operation are strong enough to paint one side of a pillar red and the other yellow.

In a few more weeks Miss Clare will drive up to the farm again and (we hope) pass our buildings for milk production. We shall be licensed to produce and sell attested Channel Island milk. It will be exciting to see the first empty churn awaiting our first consignment from the farm; though we shall have little enough to put in it until the younger cattle calve next year. But we know at least enough about cattle now, after all our house-cow years, to have no illusions about happy-ever-after. When you have cows you have a very uncertain element in your life. They may contract mastitis, with incalculable results, even in these days of penicillin treatment; they may present you with a row of useless bull calves (this has been happening for the past eighteen months, not only here, but at both Upper and Lower Highfield); they may fail to get in calf and have to be sold at a dead loss for beef; they may break out of the farm, or into the cornfields or orchards, ruining crops and young trees; they may horn one another — or you — and they may "do a swift one" on you by drying up suddenly after a promising start. With cows, you never know. But at least there is, theoretically, a fifty per cent chance that everything will work the other way round (Ridgeway and Redfields Farms had runs of heifer calves and Upper Ridgeway had twin heifers in the summer). And if, as someone said to me the other day, you are the sort of person that likes that kind of life, then that is the kind of life you like.

CHAPTER TWENTY-FIVE

Summing Up

Now, a few weeks after writing the previous pages, we are engaged in T.T. milk production. Miss Clare came back to the farm and passed our milking parlour and dairy, so that at last we are licensed producers. At this moment, I can hear the milk lorry rumbling away down the drive with its load of churns clanking; and Shelley is cheerfully banging about in the dairy where she is washing and sterilizing the utensils. Looking up from my desk, I can see the stable, not looking like its old self at all. Bill rebuilt it and floored it, I retiled it, Bill Davy plastered it and Shelley snow-cemmed the inside walls. Outside, there is the concrete "apron," made by Bill and Bill Davy, and the empty churn for tomorrow morning's milk stands on it waiting to be carried into the dairy. Next year there will be more churns, when the heifers begin to calve down.

Redfields Misty has calved; it was another bull calf — our fifth in succession. There will be no more calves now until May, when Coronet is due. Coronet had been a problem; we could not get her in calf to A.I. At the failure of the twelfth attempt we decided we would like to try her once with a bull before sending her to market as a barrener. Our vet. was not enthusiastic about the advantages of natural over artificial service. If she was able to conceive at all, he said, she would be just as likely to conceive to A.I.

The fact is that many people think differently. Both Ridgeway and Upper Highfield Farms keep bulls because they have found that the fertility rate is higher with natural service than it was with A.I., and one is constantly hearing the same tale. However, on small farms, where the keeping of a bull is impracticable, the artificial insemination service is of the greatest value, except for this one drawback.

The bull at Upper Highfield was successful with Coronet on her first and only visit. Unfortunately for us, he is a Guernsey, but it is better to have a cross-bred calf and a full lactation than no calf, no milk, and no cow at all. We should have been particularly sorry to lose Coronet, who was the first calf we reared, and is a good milker and easy to handle: the children learned on her.

When Shaun comes home for Christmas he will see great changes. I hope he will not miss the butter and cream of our house-cow days too much, and will agree with us that to have a dairy herd, no matter how small, is far better; but perhaps we shall be able to keep a cow for the house again, in the future.

Six years is a long time in the life of a child. When we came here Shaun was four and Shelley eleven. It seems no time at all to us, grown-up as we are, but Shaun can hardly remember when he did not live on a farm.

People sometimes ask us if we have any regrets. Would we recommend a similar venture to other non-farming families? This is a difficult question on which to generalize because it depends on several different factors. We, for example, were not entirely without country and farming knowledge, but this is not enough. Optimism, staying power and ingenuity . . . all these are essential. Even, however, with all these in one's favour I am, after six years, convinced that no one with children dependent upon them should undertake such a venture without an adequate secondary income or enough capital to fall back upon in those first precarious years.

Given the right temperament and, in addition, a true abiding love of the country in all its moods plus financial security, then, yes, I am all for it. I believe that there are no greater rewards than those which fall to the farming family who work hard in their own fields with their own stock. The work is, indeed, arduous, far more so than most townsfolk, with their eight-hour day and fortnight's annual holiday, can imagine. But it is interesting, healthy, rewarding and constructive. It is conducive to simple happiness and a quiet mind.

ISIS publish a wide range of books in large print, from fiction to biography. A full list of titles is available free of charge from the address below. Alternatively, contact your local library for details of their collection of ISIS books.

Details of ISIS unabridged audio books are also available.

Any suggestions for books you would like to see in large print or audio are always welcome.

ISIS
7 Centremead
Osney Mead
Oxford OX2 0ES
(0865) 250333

BIOGRAPHY AND AUTOBIOGRAPHY

David Bret	**Maurice Chevalier**
Sven Broman	**Garbo on Garbo**
Pauline Collins	**Letter to Louise**
Earl Conrad	**Errol Flynn**
Quentin Falk	**Anthony Hopkins**
Clive Fisher	**Noël Coward**
Sir John Gielgud	**Backward Glances**
Reggie Grenfell & Richard Garnett	**Joyce By Herself and Her Friends (A)**
Michael Hordern	**A World Elsewhere**
Joanna Lumley	**Stare Back and Smile**
Shirley MacLaine	**Dance While You Can**
Arthur Marshall	**Follow the Sun**
Sheriday Morley	**Robert, My Father**
Michael Munn	**Hollywood Rogues**
Peter O'Toole	**Loitering With Intent**
Adua Pavarotti	**Pavarotti**
Hilton Tims	**Once a Wicked Lady**
Peter Underwood	**Death in Hollywood**
Alexander Walker	**Elizabeth**
Aissa Wayne	**John Wayne, My Father**
Jane Ellen Wayne	**Clark Gable**
Jane Ellen Wayne	**The Life and Loves of Grace Kelly**

BIOGRAPHY AND AUTOBIOGRAPHY

Paul James	**Margaret**
Paul James	**Princess Alexandra**
John Kerr	**Queen Victoria's Scottish Diaries**
Margaret Lane	**The Tale of Beatrix Potter**
Bernard Levin	**The Way We Live Now**
Margaret Lewis	**Ngaio Marsh**
Vera Lynn	**Unsung Heroines**
Peter Medawar	**Memoir of a Thinking Radish**
Michael Nicholson	**Natasha's Story**
Angela Patmore	**Marje**
Marjorie Quarton	**Saturday's Child**
Martyn Shallcross	**The Private World of Daphne Du Maurier**
Frank and Joan Shaw	**We Remember the Blitz**
Joyce Storey	**Our Joyce**
Douglas Sutherland	**Born Yesterday**
James Whitaker	**Diana v. Charles**

(A) Large Print books also available in Audio

TRAVEL, ADVENTURE AND EXPLORATION

Jacques Cousteau	**The Silent World**
Peter Davies	**The Farms of Home**
Patrick Leigh Fermor	**Three Letters From the Andes**
Keath Fraser	**Worst Journeys**
John Hillaby	**Journey to the Gods**
Dervla Murphy	**The Ukimwi Road**
Freya Stark	**The Southern Gates of Arabia**
Tom Vernon	**Fat Man in Argentina**
A Wainwright	**Wainwright in the Limestone Dales**
Dylan Winter	**A Hack in the Borders**

(A) Large Print books also available in Audio

WORLD WAR II

Paul Brickhill	**The Dam Busters**
Reinhold Eggers	**Escape From Colditz**
Fey von Hassell	**A Mother's War**
Dorothy Brewer Kerr	**The Girls Behind the Guns**
Vera Lynn	**We'll Meet Again** (A)
Vera Lynn	**Unsung Heroines**
Tom Quinn	**Sea War**
Frank and Joan Shaw	**We Remember the Battle of Britain**
Frank and Joan Shaw	**We Remember the Blitz**
Frank and Joan Shaw	**We Remember D-Day**
William Sparks	**The Last of the Cockleshell Heroes**
Anne Valery	**Talking About the War**

POETRY

Long Remembered: Narrative Poems

INSPIRATIONAL

Thora Hird	**Thora Hird's Praise Be! Notebook**

REFERENCE AND DICTIONARIES

The Longman English Dictionary
The Longman Medical Dictionary

GENERAL NON-FICTION

Eric Delderfield	**Eric Delderfield's Bumper Book of True Animal Stories**
Caroline Elliot	**The BBC Book of Royal Memories 1947-1990**
Joan Grant	**The Cuckoo on the Kettle**
Joan Grant	**The Owl on the Teapot**
Helene Hanff	**Letters From New York**
Martin Lloyd-Elliott	**City Ablaze**
Elizabeth Longford	**Royal Throne**
Joanna Lumley	**Forces Sweethearts**
Vera Lynn	**We'll Meet Again**
Desmond Morris	**The Animal Contract**
Anne Scott-James and Osbert Lancaster	**The Pleasure Garden**
Les Stocker	**The Hedgehog and Friends**
Elisabeth Svendsen	**Down Among the Donkeys**
Gloria Wood and Paul Thompson	**The Nineties**
The Lady Wardington	**Superhints for Gardeners**
Nicholas Witchell	**The Loch Ness Story**

SHORT STORIES

Thomas Godfrey	**Country House Murders, Volume 2**
Thomas Godfrey	**Country House Murders, Volume 3**
M R James	**Ghost Stories of An Antiquary (A)**
Stephen King	**Night Shift**
Louis L'Amour	**The Outlaws of Mesquite**

HUMOUR

Douglas Adams	**Mostly Harmless**
Daphne du Maurier	**Rule Britannia**
Terry Pratchett	**Equal Rites**
David Renwick	**One Foot in the Grave**
Tom Sharpe	**Ancestral Vices**
Tom Sharpe	**The Great Pursuit**

(A) Large Print books also available in Audio